ENVIRONMENTAL
INTELLIGENCE
UNIT

# MARINE PROTECTED AREAS AND OCEAN CONSERVATION

## Tundi Spring Agardy, Ph.D.

World Wildlife Fund
Washington, D.C., U.S.A.

Academic Press

R.G. LANDES COMPANY
AUSTIN

# ENVIRONMENTAL INTELLIGENCE UNIT

## MARINE PROTECTED AREAS AND OCEAN CONSERVATION

R.G. LANDES COMPANY
Austin, Texas, U.S.A.

This book is printed on acid-free paper.
Copyright 1997 © by R.G. Landes Company and Academic Press, Inc.

Please address all inquiries to the Publisher:
R.G. Landes Company
810 S. Church Street, Georgetown, Texas, U.S.A. 78626
Phone: 512/ 863 7762; FAX: 512/ 863 0081

Academic Press, Inc.
525 B Street, Suite 1900, San Diego, California, U.S.A. 92101-4495

United Kingdom Edition published by Academic Press Limited
24-28 Oval Road, London NW1 7DX, United Kingdom

Library of Congress Catalog Number: 96-40227
International Standard Book Number (ISBN): 0-12-044455-0

While the authors, editors and publisher believe that drug selection and dosage and the specifications and usage of equipment and devices, as set forth in this book, are in accord with current recommendations and practice at the time of publication, they make no warranty, expressed or implied, with respect to material described in this book. In view of the ongoing research, equipment development, changes in governmental regulations and the rapid accumulation of information relating to the biomedical sciences, the reader is urged to carefully review and evaluate the information provided herein.

### Library of Congress Cataloging-in-Publication Data

Agardy, Tundi.
    Marine protected areas and ocean conservation / Tundi Spring Agardy
       p. cm. — (Environmental intelligence unit)
    Includes bibliographical references and index.
    ISBN 1-57059-423-6 (alk. paper)
    1. Marine parks and reserves.   2. Marine resources conservation.   3. Biological diversity
conservation.  I. Title.  II. Series.
QH91.75.A1A35  1997
333.91'6416—dc21
                                           96-40227
                                            CIP

Printed and bound by CPI Group (UK) Ltd, Croydon, CR0 4YY

Transferred to Digital Print 2011

# Publisher's Note

R.G. Landes Company publishes six book series: *Medical Intelligence Unit, Molecular Biology Intelligence Unit, Neuroscience Intelligence Unit, Tissue Engineering Intelligence Unit, Biotechnology Intelligence Unit* and *Environmental Intelligence Unit*. The authors of our books are acknowledged leaders in their fields and the topics are unique. Almost without exception, no other similar books exist on these topics.

Our goal is to publish books in important and rapidly changing areas of bioscience and the environment for sophisticated researchers and clinicians. To achieve this goal, we have accelerated our publishing program to conform to the fast pace in which information grows in the biosciences. Most of our books are published within 90 to 120 days of receipt of the manuscript. We would like to thank our readers for their continuing interest and welcome any comments or suggestions they may have for future books.

Shyamali Ghosh
Publications Director
R.G. Landes Company

# CONTENTS

Imagine a vast, largely unexplored space: effused with life, dynamic and without defined boundaries. Living things are suspended in a moving, fluid three dimensions, where even plants—the foundation for large and complex food chains—move. Note that this massive, vibrant web of life that knows no bounds cannot be claimed; it is the common property of all yet belongs to none. The space contains resources important to humanity, and supports processes critical to all of life on earth. This extensive valuable system is as complicated as a living being, with interdependent interactions and feedback mechanisms that have existed in a careful state of balance since the beginning of life itself. Yet for all its vastness and its seemingly limitless productivity, the health of this system deteriorates before your very eyes.

Now imagine that you are charged with protecting this dynamic and mysterious world—that you have been given the responsibility to control the exploitation of its resources, to coordinate the myriad uses it confers and to mitigate the indirect activities that are causing it to become stressed and degraded. However, you can only administer your responsibility in a narrow little corner of its vast domain, a minute space in a world of competing and conflicting interests. For all that you do within the limits of your charge, the continued functioning of the whole will depend on much more, on careful use and attention everywhere.

Such is the status of the world's ocean space, and it appears to be in dire trouble. Nowhere is this more true than in the fragile coastal belt of the world's continents: home now to three-quarters of the world's population and having to withstand a population growing faster than any other area. The signs that coastal degradation is indeed occurring—the fish kills, algal blooms, bleached coral reefs, decreasing commercial fish catches—confront us daily and go largely unnoticed. Yet, these small signs are like droplets of water that presage the bursting of a dam.

Faced with the seemingly insurmountable problems of demand-driven overexploitation of marine resources, conflicting uses, overlapping jurisdictions, ill-defined boundaries and the ever-growing problem of chronic yet insidious indirect degradation of nearshore habitats caused by a wide range of human activities, no one would blame marine resource managers if they threw up their hands in resignation and defeat. However, for all the vastness of problems facing marine managers, we have made great headway in recent years. The promise of a better future holds bright—if only the oceans will maintain themselves until we slowly but surely get our act together.

There are three main areas where marine conservation generally, and marine protected area planning more specifically, shows signs of revolutionary advance.

The three areas are divergent and include sociological, economic and scientific, yet the three together are necessary elements of modern and effective protected area planning.

In science-based management, advances in the fields of population dynamics, systems ecology and physical oceanography are used to identify areas that are truly critical to protection in order to keep the ecosystem functioning. This is science in the service of management; management that, because it is truly ecological in basis, is sustainable.

The field of resource economics is also starting to give marine conservation a firmer leg upon which to stand. A sudden interest in valuations has demonstrated that major incentives exist for managing coastal and marine areas wisely; justifications that are much in demand in countries that were previously faced with the choice of economic development or conservation. Through more comprehensive and sophisticated analyses, economic models can demonstrate that a healthy coastal system is a valuable asset to a nation, worthy of its protection and capable of high returns on investment.

Sociologically, the last decade has witnessed a turnaround in the way that responsibility for environmental protection and resource management is divested to local users themselves. Beyond the grandiose global schemes to protect the environment and the glutted and burdensome government administrations that oversee resource management on a large scale, the real success stories come from small scale protected areas in which stewardship for marine resources is a prevalent and inalienable facet of local coastal societies.

In land-based park planning, an area can be fenced off and buffered from negative human impact. As long as the protected area is large enough, it can by default contain enough of the essential processes to keep the system self-sustaining. To design marine protected areas in this way would be inconceivable; they cannot be fenced off, for one, and the geographically large-scale linkages of marine habitats mean that trying to envelope a self-sustaining system would require a huge unmanageable area. For truly effective marine protected areas, we must start by assessing how the system we are targeting for conservation works. By identifying what is critical to an ecosystem, in terms of nutrient loading and transport, reproductive and nursery habitats, migratory corridors, etc., planners can design a protected area that accommodates a wide variety of uses while preserving that which is most critical to ecosystem function. Scientists must then continue to provide assistance in monitoring the effectiveness of management.

In the new generation of marine protected areas, critical areas can be protected as small but highly protected "core areas." Such core areas need not be large and need not preempt continued human use of ocean space. Instead, a suite of small but strictly protected core areas will ensure that the ecosystem's self-sustaining processes are protected, guaranteeing further production and providing the basis for continued sustainable use. For marine populations with recruitment from afar, however, protection measures will have to reflect ecological reality, suggesting that core areas may have to be designated in sites far removed from the original habitat requiring conservation attention.

Multiple use marine and coastal protected areas thus make conservation and development compatible, rather than conflicting, activities. Well-managed zones of regulated use can accommodate a large range of activities if they are specifically designed to buffer critical core areas. If successful, the buffer zones then act to reduce chances that the function of critical areas is undermined by indirect impacts. Encompassing an entire set of core and buffer zones is an outer boundary delimiting an area of cooperation: a large marine ecosystem that all agencies and interested parties have agreed to jointly manage and protect.

Ultimately, decades of time will have to be the gauge by which we measure the effectiveness of the new generation of coastal or marine protected areas. We must be willing not only to help plan and implement these new multiple use reserves, but also to develop ways to monitor their progress and see how well objectives are being met. This is especially true since our knowledge of marine systems is still in its infancy and we have many mysteries to uncover and lessons yet to learn. As we live in a dynamic world, changing at a faster pace than ever, we must also prepare ourselves to respond to change, to give up old models that are based on a false, static world and to learn through trial and error as well as our occasional enduring success stories.

## Acknowledgments

As the bulk of this book was written while I was pregnant—not only with ideas but also with my daughter Alex—thanks go first and foremost to my wonderfully encouraging and good-humored husband Josh Spring. The most productive period of my life could not have happened without him! Warm thanks also to Paul Dayton (and his esteemed colleague Fran Bacon), whose insight and intellect provided the foundation for much of this writing, and to Billy Causey, who is a constant inspiration. Finally, I am much indebted to Gary Hartshorn, who first got me going as a full-fledged conservationist and whose singular support for my work at WWF is now greatly missed.

# Part I:
# Marine Conservation Issues

# CHAPTER 1

# INTRODUCTION TO MARINE SYSTEMS: BIODIVERSITY, ECOLOGY AND VALUE TO HUMANS

## SECTION 1. MARINE BIODIVERSITY AND WHY IT STANDS THREATENED

To most of us, the oceans seem bountiful and beckoning, unvarying and vast and, for the most part, mysterious yet at the same time immensely stable. The shimmering surface of the sea largely shrouds the inner workings of complex ecosystems that the oceans embody. Below the surface, however, currents, gyres, upwellings and deep ocean rivers spell unending flux. Over five hundred million cubic kilometers of water are contained in the world's oceans, every molecule moving. One cannot help but be awed by the sheer size of this dynamic world and by the thought that, for myriad organisms, sea water spells life itself.

We all know that for the bulk of living things on earth, life begins and ends and begins again in the oceans. We know this because we have been taught evolution and ecology, nutrient cycling and meteorology. But we also know this more fundamentally without being told. A primordial pull draws us to the sea and once there, we can feel–in a primitive, unquantifiable way–our ocean origins and our connection with all living beings.

Circling overhead or flitting just above the whitecaps, seabirds like the stormy petrel and the albatross are invisibly yet inevitably tethered to the workings of the open ocean system. Spray created by wind on

*Marine Protected Areas and Ocean Conservation*, by Tundi Spring Agardy.
© 1997 R.G. Landes Company.

water blasts bacteria and other organisms of the sea surface microlayer into the atmosphere, ecologically linking the two systems. Larger marine organisms also become airborne: flying fish fleeing from predators launch themselves aloft for a few precious moments of escape, only to land back in dangerous waters again.

Near the surface, where upper ocean currents meander towards one another, collide and veer off again, a thin stretch of sargassum and other weeds collect in a convergence zone. If one looks carefully among the floating weed, a bustling and gloriously diverse little community is exposed. Copepods and other crustaceans, small pelagic fishes, ctenophores and jellyfish, larval and juvenile coastal fish and even tiny hatchling sea turtles make the rich weed belt their home. Some visit temporarily, using the weed lines as a nursery habitat, while others stay their entire lives.

Below the surface, microscopic plants collectively known as phytoplankton provide the basis for a huge, hungry food chain. Considerably more photosynthesis is accomplished by these plants than by all the plants on land. Phytoplankton are food for zooplankters–small animals that live in the ocean's top photic layers and drift with the upper currents. Phytoplankton and zooplankton themselves also provide sustenance for filter feeding fish and mammals–animals as large as the earth's largest, the blue whale.

Light from the overhead sun is largely refracted at the sea's crinkled surface, but some light does indeed penetrate the dense seawater, travelling down to depths of 200 meters. In the blue-green tinged water of the upper oceans, light from the sun breaks into thick shafts that deliver energy to seemingly impenetrable and dark voids below. The angled beams of light give the underwater scene a sacred, ethereal look somehow.

Travelling down from the surface, one encounters blooms of single-celled algae, diatoms, isopods and copepods, moon jellies and medusoid jellyfish. Synchronized schools of pelagic hunters, the mackerel, bonito, or sub-adult albacore, move like silent amorphous ghosts into and then out of view. Occasionally, a lone predatory white marlin or mako shark follows on the schools' heels.

Down where the water turns an ominous blue-black and the light shafts end in sword-like points that cannot pierce the bottom blackness, midwater fishes swim and rest. Some of these midwater species practice diel migration, rising towards the surface during the safety of night, then vertically migrating downwards with the first tenuous signs of daylight above. Eels pass through on their way from the distant river mouths where they spent their early lives to the mid-ocean spawning grounds where, once spent, they will die. Giant shadows are cast, even in this dimness, when a sperm whale dives down into the depths, scattering fish and invertebrates in its wake. Not far away but visually hidden by the blanket of darkness, a giant squid senses the sperm whale hunting for it and propels itself even deeper into the abyss.

Where blackness rules, life still continues. The depths of the open sea are teaming with thousands of species, each adapted to the rigors of living without light in its own peculiar way. Bizarre looking flash-light fishes blink signals to each other through ink-black seawater, communicating who they are (species identification) and what they want (courtship) with frenzy. Sometimes the thick darkness seems empty but then an eerie glow appears in the distance. If you go right up to it, the looming, glowing thunderhead shape reveals itself as a dense congregation of jellyfish, each emitting cold light known as biolumi-nescence. Leatherback sea turtles that weigh a thousand kilograms and are shaped like depth-sounding torpedoes search the depths for these jellyfish aggregations upon which they feed.

If you are travelling down through the deep sea with searchlights, nearing the bottom is startling. The benthos looms suddenly, a shock of form and life. On the surface of the seafloor, resembling nothing more than the pock-marked moonscape, things are on the move. Able to withstand darkness, cold and immense pressure, the deepsea world, even five kilometers down, is pulsating with life. The batfish glide slowly while the chimeras scurry and those invertebrates that have the luxury of mobility creep about avoiding predation. The trick to life down here is being lucky, for many of the creatures that live in the deep blackness must depend on rare encounters with prey or carrion for sustenance.

Farther along the seafloor, moving slowly away from the direction of the distant shore, awaits the biggest surprise of all. A long fissure in the seabed seems devoid of life where it begins as a small crack in the muddy bottom. As one moves along the widening crevice to the place where thick plumes of what look like coal plant smoke bellow from the center of the earth itself. Here, suddenly, unexpectedly, without logic, springs to life a colorful, odd, diverse assemblage of life forms. Giant mussels, sea feathers and tube worms crowd around the vent opening. We might as well be on the surface of the moon because these life forms have laughed in the face of nature's laws: they live without connection to all other living things since they derive no energy, directly or indirectly, from the sun. These alien species depend on chemosynthetic bacteria that harness the sulphurous energy of undersea volcanos and the distant sun doesn't exert any influence.

However, taking an imaginary journey to the deep ocean depths is not the only way to be impressed by the abundance, diversity and strangeness of life in the sea. Even close to shore, in waters that we mistakenly think of as well-studied and known, mysteries abound. We know precious little of what is out there and how the marine ecosystem functions. Pity us poor land creatures, for whom ocean study and exploration is difficult and largely out of reach.

What we do know is staggering. Oceans harbor far more biological diversity than land, however, the biodiversity is concentrated not at the species level but higher up in the taxonomic order. There are a total of 32 known major phyla in the seas and of these, 15 are exclusively marine. Compare this to only one phylum that is confined to land. Many more phyla, classes and genera of organisms means a virtual explosion of life forms–dramatic differences in size and shape and means of living abound.

Why is the tapestry of ocean life so rich? What makes this remarkable biological diversity possible? One answer has to do with the fluid that comprises over seventy percent of the planet's surface. In a wide variety of ways, water is the most hospitable medium for life. Its unusually high capacity for storing heat makes it a buffer against the vagaries of the climate, its exceptional ability to dissolve substances provides an environment in which chemicals needed for life support are readily available, its fluid nature allows metabolic wastes to be disposed of easily and its density means that body forms can afford to be unconventional since constraints imposed by gravity are greatly relaxed. Finally, water itself is essential to all life and this water ultimately comes from the oceans. It is, thus, no coincidence that blood and other organic fluids have ionic compositions mirroring seawater. Even with our terrestrial way of life, we carry a bit of our ocean heritage around with us.

So, the oceans are natural for supporting the origin of life and its subsequent gaudy expression into the riot of marine life forms we know today. But, paradoxically, the marine environment also presents challenges for survival. The vastness of the oceans and the sheer power of its movements makes it hard for any living thing to be in the right place at the right time. In a constantly changing world, chance plays a big role in survival. It is thus this combination of plentiful resources in an environment conducive to life, yet conditions like low light, patchily distributed nutrients, or strong currents that challenge emerging life forms that has led to the profusion of highly specialized marine species. And time, too, is a factor; the oceans have been the site of experiments in nature for over three-and-a-half billion years.

Incredibly rich though the oceans are, when people think about marine diversity they usually think only of coral reefs. But all marine habitats exhibit a wonderful cadre of organisms if we take the time to observe them. The rocky intertidal community of temperate coasts, for instance, displays a wide variety of animals and plants specially adapted to living in the harsh environment of the wave-pounded tidal zone. Here sea stars, purple urchins, brittle stars and other echinoderms cling fiercely to rocks, competing for footholds (they each have hundreds of tube feet to accommodate) with chitons, limpets and other primitive molluscs. Mussels and other pelycopods cram the rock crevices and a wide array of seaweed species grow anywhere they are able.

Crabs and other crustaceans clamber about fleet of foot, avoiding predatory gulls while searching for their own food.

Many factors are thought to contribute to the maintenance of these diverse, interconnected communities. They clearly depend on pelagic production and wave action to bathe the attached species with nutrients and spread larval propagules. The success of recruitment depends on many types of disturbance, predation and competition. In one locality a starfish species plays a key role by consuming a dominant competitor for space, freeing resources for many other species to thrive.

In nearshore temperate zones, the kelp forests rival the tropical rainforests in complexity and productivity. Laminarians like the giant and elk kelps are the "trees," fish and marine mammals the "birds" of the undersea forest. Like terrestrial forests, kelp habitats show community stratification. Invertebrates are everywhere, bryozoans colonize the kelp blades with their lace-like structures, molluscs like the abalone swim about and provide a rich food source for sea otters, and sea hares (nudibranchs) graze the canopy or the seafloor, depending on the species. Fishes include the gaudy garibaldi, the slow-growing sheepshead and the decidedly ugly cabezon. Even pelagic species like the great white shark are quick to capitalize on the habitat's vast productivity during feeding visits to the kelp forest.

Another type of marine forest is one which looks markedly terrestrial, until one gets close enough to notice that all the root systems are in fact submerged. Mangrove forests are critically important to the tropical coastal and marine ecosystems, as they act to accrete land, filter water of pollutants, produce oxygen and supply nutrients to the marine ecosystem and provide habitat for young coastal as well as pelagic animals. A snorkeling trip among the mangrove will yield juvenile and sub-adult species of coral reef fishes, sharks, snook, mullet and even tarpon. Coastal bird species like the frigate bird, booby, pelican and herons utilize the mangrove for nesting and roosting, while other bird species forage among the extensive root systems. Species specialized for life in the mangrove include mangrove oysters, mangrove snapper and other fishes and a whole host of shrimp, decorator crabs, annelid (segmented) worms, tunicates, sponges, jellyfish and algae. Without mangrove and its biological richness, many other linked habitats would be unable to support the number of species they do and would be nowhere near as productive.

Other surprising "hotspots" for marine biodiversity are the temperate, subtropical and tropical mudflats found worldwide. A cursory glance at one of these habitats would undoubtedly leave a viewer unimpressed, but beneath the ooze and between the particles of mud and sand thrive a multitude of uniquely adapted species. In the Gulf of Maine region of Canada, for instance, over a thousand invertebrate species were found to inhabit one small bay. Gastropods like the moon snails and the variably patterned snails plow through the silt, clams settle into the mud with only their siphons exposed and different crab species variously live in, under, or beside the flats. Mudflats also provide vital resources for transient visitors, like migratory shorebirds who know about the cornucopia that lies beneath the mud's surface and industriously forage for it.

Then there are the coral reefs and seagrass meadows, glorified in travel brochures and television documentaries. Rich in color, form and species and a virtual catalogue of specialization, the coral reefs compete with tropical forests for our attention and concern. Unfortunately, we seem to be loving them to death. Coral reef communities are extremely sensitive to environmental change and disturbance, having a narrow range of tolerance for salinity, temperature, light and water quality. Changes in tropical marine environments brought about by natural forces and human-induced stresses (inadvertently caused by our fishing, scuba diving, glass bottom boating and swimming) have reduced the genetic and species diversity of coral reefs and associated habitats. This, combined with geographically large scale effects like pollution and increasing temperatures that result in coral bleaching, has acted to undermine coral reef and other nearshore tropical systems around the world.

We might be tempted to compartmentalize the seas, to say one area ranks over another because it has more species, shows more endemism or unique specialization, or is more appealing. We sometimes speak of marine systems like coral reefs and mangrove forests and hydrothermal vents as if they were rooms in a living museum—each room with its own character and set of species on display—static, immutable and free standing. We would be wrong to do so. The great web of life is nowhere stronger in its connections than in the sea and all habitats and the communities they contain, are ultimately linked. Changes occurring in one corner of the marine realm have effects throughout. As we peer out from our comfortable shores at an unfathomable and mysterious ocean, we would do well to remember that we terrestrial beings, too, are influenced and nourished by the seas in ways we can only begin to understand.

The coastal plains, nearshore waters and open ocean areas supply human beings around the globe with vital supplies such as food and minerals, transportation and energy, places for recreation and tourism. These biomes also harbor the majority of the ecological processes critical to the functioning of the biosphere. Despite this and although the oceans cover nearly three-quarters of the earth's surface, protection of the ocean environment and its processes appears at the bottom of most lists of conservation priorities.[1]

Coastal and marine areas are certainly beginning to get the conservation attention they need and deserve, but are afforded nowhere near the protection needed to conserve them in sustainable fashion over the long-term. Marine systems form the largest set of biomes on the planet. The coastal margins of the continents currently support most of the world's human population and will absorb most of its

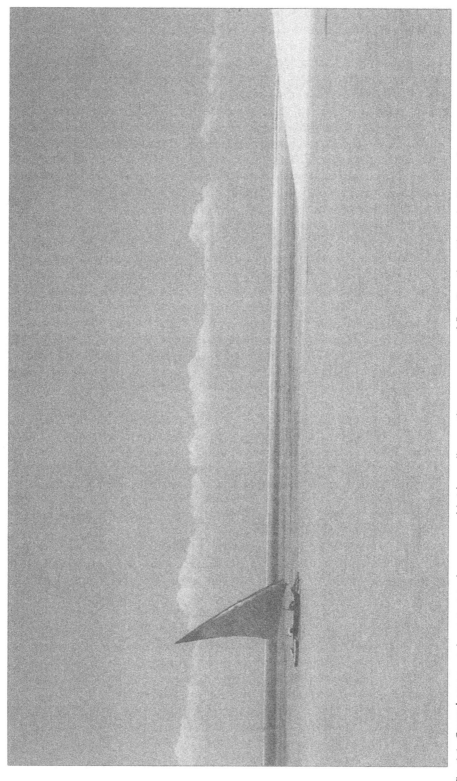

*Fig. 1.1. Coastal areas are important to humans worldwide: a dhow sailing across a reef flat, Zanzibar. Photo by T. Agardy.*

increase when the global population approaches 8.5 billion in the next millennium. Nearshore waters support the world's most naturally productive ecosystems: estuaries and lagoons, brackish water wetlands, mudflats and other intertidal areas, coral reefs and seagrass beds. By some measures the oceans support the highest biological diversity in the world, rivalling that of tropical forests and exceeding all other areas in diversity of genera, classes and phyla.[2] These habitats are critically important to mankind, containing the best alluvial soils for agriculture and the most productive fishing grounds and supporting water-dependent and enhanced industry, tourism and waste disposal.[3] All in all, marine ecosystems are complex, ecologically sensitive and exceedingly valuable places that are under enormous, and in most cases, largely ignored threats.

The burgeoning human population and the stresses it exerts on the ecosystem, inequitable distributions of global wealth and inefficient consumption practices all contribute to continued environmental degradation. This is particularly true in the world's coastal zones, which by the year 2000 will accommodate nearly three-quarters of humanity. Our track record at managing our negative impacts on the coastal and near-shore marine environment is not laudable even now and will only be worse as resource use conflicts increase. Meanwhile, the vast oceans are becoming despoiled and the very processes on which most life depends are put at risk.[4]

Much of the world perceives the oceans as homogeneous, resilient and so vast as to be essentially limitless. Yet, coastal areas have become a giant sink for our land-based and ocean-dumped wastes and our naive beliefs in the endless assimilative capacity of the seas has led to a collective "out of sight, out of mind" philosophy. While we have come to recognize the spectacular diversity and importance of habitats like tropical forests, we note that the species diversity of the oceans appears to pale in comparison.[5] In effect, we think of the oceans as being neither particularly valuable to us, nor particularly threatened. But we could not be more mistaken.

With a few notable exceptions, the public at large and decision-makers who represent them have resisted major involvement with marine conservation issues. In part this has to do with this "out of sight, out of mind" phenomenon and the erroneous perception that the vast oceans have a nearly infinite capacity to assimilate wastes and resist degradation. It also has to do, of course, with how little we know about ocean ecology and how expensive it is to gain more knowledge. Yet we do know enough to be sure that chronic assaults on coastal and ocean areas, even those that appear trivial and inconsequential, are having dramatic negative impacts on the marine system. Fish kills, massive algal blooms, loss of biodiversity, extirpations of species and populations and increased frequency of cataclysmic meteorological events are only some of the warning signs.

Marine conservation that is comprehensive, integrative and caters to human needs as well as environmental concerns is urgently needed around the globe. Marine protected areas–areas specially managed to protect species, the habitats that support them and the ecosystems that they comprise–are a key tool in saving the earth's seas from ecological ruin. Marine protected areas serve to achieve three goals simultaneously: 1) they can be used to protect marine and coastal biological diversity (the variety of life at all levels that exists in coastal zones and open ocean areas); 2) they can ensure that marine productivity is not undermined by uncontrolled exploitation (Dayton et al, 1995); and 3) they can be used to focus resources and energies towards restoration of vital areas that are already degraded yet have good potential to support the marine ecosystem and human beings in the future.[6]

Marine biological diversity–encompassing the enormous variety of marine species, the cornucopia of living marine resources, myriad coastal and open ocean habitats and the wealth of ecological processes that support them–is something that we are only now beginning to appreciate as being vital to all of our lives.

The world is becoming increasingly aware of the mutual interdependence between the quality of human life and the condition of the environment that sustains and nurtures us in countless ways.[7] While most nation states have signalled their interest in protecting while sustainably using their terrestrial ecosystems, as witnessed by numerous international agreements from the Earth Summit to the Convention on Biological Diversity, the same degree of awareness and commitment to conservation has not been afforded coastal and ocean areas under their national jurisdictions. Now, even as we recognize the danger in neglecting marine conservation, the seas are beginning to show signs of serious and often irreversible, degradation. If we turn our backs on the world's oceans and their ecosystems, or even if we end up doing too little too late, we stand to lose more than important stocks of resources and opportunities for coastal development. We will have undermined the very life-support system of the globe that makes life on the planet possible.[8]

Coastal and oceanic systems harbor far more biological diversity than terrestrial ecosystems, although diversity is not concentrated at the species level but instead occurs at both the population level (as genetic diversity) and at higher levels of taxa. As stated previously, 32 of 33 known phyla occur in the seas and 15 of these are exclusively marine.[5] The range of adaptation of marine creatures is staggering. While sea creatures exhibit virtually the entire range of existing basic body plans, terrestrial and freshwater animals comprise many variations on relatively few themes in those few phyla with enormous numbers of species (notably insects, nematodes and annelids). Marine biological diversity is notable at the mega scale as well. Ecological communities in the coastal and marine realms are generally more complex and cover far wider geographical scales in terrestrial biomes, with the presence of marine filter-feeders adding an extra trophic layer to the food web.[9,10] Lastly, diversity of size is significantly greater in the seas, with organisms ranging in size from picoplankton to the great whales.

Open ocean and coastal habitat types run the gamut from temperate and tropical wetlands such as saltmarshes and mangrove forests, to rocky intertidal areas and sandy beaches, through mudflats and seagrass meadows, to coral reefs, sea mounts, hydrothermal vent systems and abyssal plains and the various types of water column communities.[11] Figure 1.2 shows the distribution of main nearshore marine ecoregions that form Large Marine Ecosystems (LMEs) worldwide. The range of ecological services that these provide in terms of storing and cycling nutrients, regulating water balances, buffering land from storms, filtering pollutants, etc., as well as the importance of the role that oceans play in regulating planetary balances in hydrology and climate, far surpasses that of land. There is an indisputable link between the biological processes occurring in the oceans and the regulation of global balances. For instance, the photosynthetic pump that removes the primary greenhouse gas, carbon dioxide, from the atmosphere contributes to the earth's climate regulating function. In so doing, this biological pump produces one-third to one-half of the global oxygen supply. The health and stability of the food web that maintains this pump must be protected in worldwide efforts to conserve biodiversity and use resources sustainably.

Unfortunately, for coastal nations that depend directly on healthy marine ecosystems for goods and services, direct and indirect degradation of nearshore waters has begun to exact a toll in the form of lowered productivity, increasing user conflict and impairment of the very ecological functions that humans value. Effects are most often cumulative over time and multifaceted in nature, such that even coastal communities are unaware of the magnitude of the damage and the loss of biodiversity until years of damaging impacts have accumulated. Not only are the oceans suffering the death of a thousand cuts, but it is ironically the most productive, most ecologically critical and most highly valued habitats that we are degrading the fastest. The nearshore estuarine habitats, intertidal areas, banks of the continental shelves and reef systems are rapidly being depleted, altered, or poisoned in permanent ways.

To date, few nations have indicated that they recognize the need to develop and implement comprehensive conservation programs aimed at protecting marine ecosystems and the species they support and even fewer have committed to realizing such goals. Attempts to conserve marine biological diversity, where they spring up, derive largely from crisis management needs. An oil spill here, a fisheries stock collapse

*Fig. 1.2. (opposite page) Distribution of main nearshore marine ecoregions by Large Marine Ecosystem delimitation (from Sherman, 1991).*

*Fig. 1.2. Large marine ecosystems. 1 Eastern Bering Sea, 2 Gulf of Alaska, 3 California Current, 4 Gulf of Mexico, 6 Southeast US Continental Shelf, 7 Northeast US Continental Shelf, 8 Scotian Shelf, 9 Newfoundland Shelf, 10 West Greenland Shelf, 11 Insular Pacific-Hawaiian, 12 Carribean Sea, 13 Humboldt Current, 14 Patagonian Shelf, 15 Brazil Current, 16 Northeast Brazil Shelf, 17 East Greenland Shelf, 18 Iceland Shelf, 19 Barents Sea, 20 Norwegian Shelf, 21 North Sea, 22 Baltic Sea, 23 Celtic-Biscay Shelf, 24 Iberian Coastal, 25 Mediterranean Sea, 26 Black Sea, 27 Canary Current, 28 Guinea Current, 29 Benguela Current, 30 Agulhas Current, 31 Somali Coastal Current, 32 Arabian Sea, 33 Red Sea, 34 Bay of Bengal, 35 South China Sea, 36 Sulu-Celebes Sea, 37 Indonesian Seas, 38 Northern Australian Shelf, 39 Great Barrier Reef, 40 New Zealand Shelf, 41 East China Sea, 42 Yellow Sea, 43 Kiroshi Current, 44 Sea of Japan, 45 Oyashio Current, 46 Sea of Okhotsk, 47 West Bering Sea, 48 Faroe Plateau, 49 Antarctic. Source: after Sherman K. (1994) and in literature.*

there, a massive algal bloom choking off coastal waters and ruining a season of beach tourism are the problems that galvanize attention and receive the spot fixes. But rather than addressing symptoms and merely applying bandages, countries would greatly benefit from proactive planning that attacks the root causes of problems. Marine protected areas will prove to be invaluable tools in these endeavors. Indeed, this is the only path to a sustainable future.

As terrestrial beings, humans have come to regard coastal and marine systems as somehow apart from and less valuable than, the land areas that we live in and have developed.[12] This is probably less true for coastal nations that derive large parts of their economies from the exploitation of marine resources , nevertheless, an innate bias that emphasizes the importance of land and its resources and ignores the importance of the seas is common worldwide. It is no wonder that effective management of marine resources and conservation of marine ecosystems that supply those resources lags far behind that of land-based conservation and sustainable use. As a result, we suffer the consequences not only of misuse and overuse of precious marine resources, but also of degradation of the biosphere brought about because of our ignorance concerning the irrefutable linkage between the health of the oceans and that of the rest of our planet.

## SECTION 2. DIFFERENCES BETWEEN MARINE AND TERRESTRIAL ECOSYSTEMS

Marine and terrestrial systems exhibit differences in scale and process and thus require somewhat different paradigms for conservation and resource management.[13] This dichotomy in management approaches can be traced to two factors: 1) the actual nature of marine and coastal systems differs markedly from that of the land (Pimm, 1989); and 2) our perceptions of ocean and land spaces is fundamentally disparate.[14,15] That is, not only are there differences in the spatial and temporal scales that define the terrestrial and marine systems, but there are also cultural biases that drive us to think of ocean systems (and sets of exploitable marine resources) differently from the way we think of terrestrial systems. Regardless of whether the differences are of kind or of degree, we must approach the management of marine areas in innovative and perhaps unique ways.

Although both terrestrial and marine systems exist in three-dimensional space (four-dimensional time-space), land-based systems are predominantly two-dimensional, with most ecological communities "rooted" to the two-dimensional plane that is the earth's surface. The sea presents quite another picture with the bulk of life moving about in a vast nonhomogeneous space and few interactions anchoring the water column to the benthos.[16-18] The sea (and its coastal interfaces) speaks of movement: transport of nutrients occurs over large distances and both passive movements and active migrations contribute to a state of flux.[19] It is not just that marine organisms can utilize the greater density of water as compared to air to suspend themselves in a medium above the substrate; it is also (more importantly) that the plants that act as the basis for an enormous and complex food web also move and fluctuate with time.[20,21] Physical features of the marine ecosystem dictate its ecological character, more so than on land.[22-25]

These are fundamental differences between terrestrial ecosystems and coastal/marine ecosystems that must be recognized before effective management or conservation can be put in place.[20,21] The ecological dichotomies fall into four categories. First, marine ecosystems are relatively poorly understood, having a much shorter history of study and presenting logistic difficulties for researchers.[12] Marine research generally cost orders of magnitude more than similar research on land and travel to study sites is complicated and costly. For the purposes of long-term study, it is difficult to create fixed sites for monitoring that will be protected over time from direct and indirect degradation or alteration by human activity.[26,27] The fact that humans are terrestrial beings also complicates the picture, as difficult conditions and complicated logistics compound even simple underwater studies.

Second, marine and coastal systems have nebulous boundaries; both in terms of the outer bounds of ecosystems and in terms of definable limits of ecological communities (guilds of organisms or biologically defined habitats) and population structure.[28] Much more so than on land, the marine environment occurs in a highly dynamic, three-dimensional form.[29] The fluid nature of the marine environment means that on a spatial scale, abiotic and biotic linkages occur across great distances, several orders of magnitude more than most terrestrial linkages.[30] Temporal scales, however, are greatly reduced in comparison to terrestrial systems, meaning the marine systems will exhibit greater rates of change and dynamism per unit time.[31,32] The movement of the water column and its relative independence from the benthic ecology means that entire suites of organisms (habitats based on communities of organisms or guilds of species) move in sometimes unpredictable ways.[33] In terms of population dynamics, we have a very limited understanding of what constitutes a population for most marine organisms and how those populations are sustained by recruitment from afar.[34,35] This means that not only are marine ecosystems highly dynamic in space and time, but their constituent populations of species cannot be described and identification of appropriate units for management is difficult.

A third factor, alluded to above, is that marine systems exhibit enormous geographical and spatial scales, yet relatively fine temporal scales.[36,37] Although processes may be connected over wide distances, environmentally induced changes are common and almost immediately discernable.[16] Finding portions of ocean ecosystems that are out of reach of human impact is nearly impossible, so control areas are often unavailable for scientific research.[15]

Finally, marine systems are driven by largely changeable and un-predictable physical processes to a much greater extent than on land. The non-linear character of most marine functions and interactions makes understanding difficult and impedes our ability to accurately predict the consequences of manipulative experiments, or more im-portantly, management.[38] Similarly, marine systems have largely un-structured food webs, so impacts of resource removal on food chain dynamics is not easy to predict.[39-41]

We also harbor a bias towards the oceans, one that may have its genesis in the easily rationalized unease we feel when we as perfectly adapted land creatures venture into what is for us a foreign and dan-gerous medium. The difficulty and expense of ocean research and our slowness in acquiring information about the oceans feeds this bias—what is unknown is to be feared. The mysteries fuel another misconcep-tion as well: the oceans are so vast and their resources so limitless, that no matter what we do to them they are bound to recover. Para-doxically, we simultaneously harbor a sense of familiarity about the sea—an almost mystical feeling that it is a part of us and a common thread among a disparate collection of coastal peoples around the world. The seas cannot be owned, they are the quintessential common prop-erty.[42] Unfortunately, this perception leads to a prevalent and poten-tially destructive feeling that because the oceans are commons, each man is as free as the next to overexploit and misuse them.

## SECTION 3. THE VALUE OF MARINE SYSTEMS TO MANKIND

Arguments about the value of biological diversity and what steps nations should take to protect it often become bogged down in ill-defined terms and unclear semantics. Individual resources, discreet stocks and species themselves each have value; market value for pecuniary goods and other less definable values for non-pecuniary goods. How-ever, biodiversity conservation in fact requires more than conventional resource management; something more than the additive efforts to manage use of resources and preserve threatened species. Conservation of bio-logical diversity must recognize that the whole is more than the sum of its parts and that the systems that have allowed biological diversity to evolve and flourish are the fundamental targets for protection. It is entire suites of organisms, with the intricate interconnections and linkages to the abiotic features of the earth's environment, that we strive to protect through conservation of biological diversity.

The present loss in marine biological diversity, like that on land, is unprecedented and alarming. It is generally agreed that such bio-logical diversity loss, the attendant decimation of stocks of living re-sources, widespread appearance of ecosystem imbalances and the im-pairment of ecological processes may well undermine the adaptive potential of these systems and their ability to meet human needs in the future.[43] This carries with it an enormous cost, far outweighing the opportunity costs associated with the loss of economic potential due to diminished individual resources.

The marginal ocean environments—coasts, bays and estuaries—are the richest portion of the marine complex. They account for over 30% of marine productivity, though only 10% of its surface and less than 1% of its volume.[44] Estuaries are, in fact, one of the richest biomes on the planet due to high carbon fixing rates and nutrient loading. Other nearshore habitats, such as coral reef systems, are notable because of their immense diversity, despite being present in relatively nutrient-poor waters.

These nearshore areas bear the brunt of our destructive actions, yet they provide most of the resources that humans so greatly value. Of course this degradation is most often chronic and cumulative, such that we are often unaware of the magnitude of the damage and the impairment of ecological potential until years of damaging impacts have accumulated. Not only are the oceans becoming less diverse and productive, they are becoming less able to support humankind even as we rely on coastal resources with ever greater intensity.[45]

Despite this degradation, the world's seas continue to provide invaluable benefits to human beings all over the world. The marine ecosystem, including the entire spectrum of coastal and ocean biomes and the linkages between them and inland habitats, is still spectacularly rich in biota and in what it offers to mankind in goods and services.[12] We still have the opportunity to stem the tide of destruction and ensure that resources will be there for our future use, if we act strategically and soon.

Methods for determining natural resource values are notoriously ineffective, given that services (including ecological processes that maintain homeostatic balances around the globe) are not sold in markets.[46] Without estimates of value, it is difficult to emphasize what costs will be incurred through the loss of marine biodiversity and therefore difficult to present the necessary justifications for expenditure of human and fiscal resources to prevent that loss. Fortunately, resource economists are applying cutting edge models to better estimate costs and benefits of biodiversity use and conservation. Beyond estimates of total market values for all marine goods traded, economists can estimate costs of resource or habitat replacement (Hardin, 1991), derive estimates of value from surrogate markets (e.g., analyzing how waterfront land prices are related to pristineness or water quality), or conduct surveys of willingness to pay.[47] However, using all of these methods in tandem may still not approximate true values of biological diversity.

It is even more difficult to determine the values of marine resources and the systems that generate them than to determine similar values for terrestrial resources. Market values of fisheries commodities provide one of the few tangible measures of marine resource value. Marine fisheries provided about 84 million tons of food to humans and livestock in 1993, and in the developing countries alone this resulted in $11 billion in revenues.[7] More and more people around the world are dependent on fisheries-derived sources of protein and current estimates suggest that fish provide the primary source of protein for over 1 billion people. However, the sea provides much more to humans than merely protein. Other marine products that can be valued using conventional market mechanisms include marine organisms harvested for the curio trade, aquarium fishes and corals for the pet trade, mangrove wood for construction, seaweeds and other algae for use as food additives and fiber, medicines derived from marine organisms, etc. Less precise estimates of value are made for goods and services that coastal and marine systems provide to the tourism and recreation industries worldwide. Ocean space has intrinsic value as well even when resources are not used–aesthetic value and amenity value.[15] Moreover, the ecological services that the seas provide, in terms of buffering land from storms, stabilizing coasts, contributing to global nutrient cycling and other planetary processes is almost impossible to value by traditional means, yet these phenomena, crucial to human survival, are, therefore, perhaps invaluable.

How can marine ecosystems be maintained so that humans and the rest of the biosphere can continue to derive value from them? In general terms, the conservation of biological diversity, in order to be effective and lasting, encompasses three complementary and necessary measures. The first is preservation of ecological processes and protection of threatened populations of organisms, species and habitats. Often this effort is termed conservation pure and simple, but it is merely one step in the pathway to effective biodiversity conservation. The second measure is determining levels of use of resources that are sustainable and undertaking management to keep within those limits. The third is fair and equitable sharing of benefits from effective management and conservation so that people who use and depend on the resources can be rewarded for use that sustains the resource base and does not impair the ecological processes that maintain it.

In the marine realm, these three measures can be ensured through rational and cost-effective policies promoting protection of vulnerable habitats and threatened species, development of fisheries policies (both national and through multilateral agreements) to guarantee that resource use is kept within sustainable limits and investment in projects and programs that promote fairer revenue sharing to benefit those who are responsible for stewardship over the resource base. As will be demonstrated later in this book, marine protected areas are the most viable and efficient tool we have available to us to ensure that marine systems are conserved for the benefit of human beings, the biosphere and the future.

REFERENCES

1. Agardy, T. Advances in marine conservation: the role of protected areas. Trends in Ecology and Evolution 1994; 9(7):2676-270.

2. Grassle, J.F. and N. Maciolek. Deep-sea species richness: regional and local diversity estimates from quantitative bottom samples. Am. Nat. 1992; 139(2):313-341.

3. Weber, P. Abandoned Seas: Reversing the Decline of the Oceans. Worldwatch Paper 166, Wash., DC, Worldwatch Institute, 1993.

4. Cherfas, J. The fringe of the ocean–under siege from land. Science 1990; 248.

5. Norse, E. Global Marine Biological Diversity. Washington, DC, Island Press, 1993.

6. Dayton, P., R. Hofman, S. Thrush and T. Agardy. Environmental effects of fishing. Aquatic Conservation: Marine and Freshwater Ecosystems 1995; 5:205-232.

7. de Fontaubert, C., D. Downes and T.S. Agardy. Protecting Marine and Coastal Biodiversity and Living Resources under the Convention on Biological Diversity. Washington, DC, Center for International Environmental Law, World Wildlife Fund and IUCN Publication, Washington, DC, 1996.

8. diCastri, F. and T. Younes. Ecosystem function of biological diversity. Biology International Special Issue 1989; 22.

9. Conrad, M. Statistical and hierarchical aspects of biological organization. In: C.H. Waddington, ed. Towards a Theoretical Biology. Edinburgh University Press, Edinburgh, 1972.

10. Steele, J.H. The Structure of Marine Ecosystems. Cambridge, MA, Harvard University Press, 1974.

11. Parsons, T.R., M. Takahashi and B. Hargrave. Biological Oceanographic Processes. Third edition. Oxford, Pergamon Press, 1984.

12. Earle, S. Sea Change. New York, G.P. Putnam's Sons, 1995.

13. Steele, J.H. A comparison of terrestrial and marine ecological systems. Nature 1985; 313:355-358.

14. Pimm, S.L. Communities oceans apart? Nature 1989; 339:13.

15. Kenchington, R.A. and M.T. Agardy. Achieving marine conservation through biosphere reserve planning. Env. Cons. 1990; 17(1): 39-44.

16. Dayton, P.K. Scaling, disturbance and dynamics: Stability of benthic marine communities. In: T. Agardy, ed. The Science of Conservation in the Coastal Zone. Proceedings of the IVth World Conference on Parks and Protected Areas. 1993. IUCN Marine Conservation & Development Report, Gland, Switzerland.

17. Nihoul, J.C. and S. Djenidi. Perspectives in three-dimensional modelling of the marine system. Elsevier Oceanogr. Ser. 1987; 45.

18. Strathmann, R.R. Why life histories evolve differently in the sea. Amer. Zool. 1990; 30:197-207.

19. Mann, K.H. and J.R. Lazier. Dynamics of Marine Ecosystems. Oxford, Blackwell Scientific Publications, 1991.

20. Bakun, A. Definition of environmental variability affecting biological processes in large marine ecosystems. In: K. Sherman and L. Herander (eds.) Variability and Management of Large Marine Ecosystems, Washington, DC AAAS Press, 1986: 89-107.

21. Frontier, S. Diversity and structure in aquatic systems. Oceanogr. Mar. Biol. Ann. Rev. 1985; 23:253-312.

22. Mann, K.H. Ecology of Coastal Waters: A Systems Approach. University of California Press, Berkeley, 1982.

23. Pimm, S.L., J.H. Lawton and J.E. Cohen. Food web patterns and their consequences. Nature 1991; 350:669-674.

24. Smith, P.E. Biological effects of ocean variability: time and space scales of biological response. Rapp. Cons. Int. Explor. Mer. 1978; 173: 117-127.

25. Starr, M., J.H. Himmelman, J.C. Therriault. Direct coupling of marine invertebrate spawning with phytoplankton blooms. Science 1991; 247:1071-1074.

26. Agardy, T. The Science of Conservation in the Coastal Zone: New Insights on How to Design, Implement and Monitor Marine Protected Areas. Proc. of the World Parks Congress 8-21 Feb. 1992, Caracas, Venezuela. IUCN, Gland, Switzerland. 1995.

27. Hayden, B.P., R.D. Dueser, J.T. Callahan and H.H. Shugart. Long term research at the Virginia Coast Reserve. Bioscience 1991; 41(5):310-325.

28. Pernetta, J. and D. Elder. Cross-sectoral, Integrated Coastal Area Planning: Guidelines and Principles for Coastal Area Development. IUCN Marine Conservation and Development Report. Gland, Switzerland, 1993.

29. Parrish, R., A. Bakun, D. Husby and C. Nelson. Comparative climatology of selected environmental processes in relation to eastern boundary current pelagic fish reproduction.In: G. Sharp and J. Csirke (eds.) Proc. of Expert Consultation to Examine Changes in Abundance and Species of Neritic Fish Resources. FAO Fish. Rep. 1983; 291(3):731-778.

30. Havens, K. Scale and structure in natural food webs. Science 1992; 257:1107-1109.

31. Lawton, J.H. Feeble links in food webs. Nature 1992; 355:19-20.

32. Steele, J.H. Marine functional diversity. Bioscience 1991; 41(7):470-474.

33. Holme, N.A. Fluctuations in the benthos of the western English Channel. Oceanologica Acta SP, 1983: 121-124.

34. Denny, M.W. and M.F. Shibata. Consequences of surf-zone turbulence for settlement and external fertilization. Am. Nat. 1989; 117:838-840.

35. James, M.K., I.J. Dight and J.C. Day. Application of larval dispersal models to zoning of the Great Barrier Reef Marine Park. Proc. of PACON 90, 16-20 July 1990 Tokyo.

36. Longhurst, A.R. Analysis of Marine Ecosystems. New York, Academic Press, 1981.

37. Dayton, P.K. The structure and regulation of some South American kelp communities. Ecol. Monogr. 1985; 55:447-468.

38. Sherman, K. The large marine ecosystem concept: research and management strategy for living marine resources. Ecol. Appl. 1991; 1(4):349-360.

39. Gulland, J.A. Food chain studies and some problems in world fisheries. In: J. Steele, ed. Marine Food Chains. Edinburgh, Oliver and Boyd Press, 1970; 296-315.

40. Gulland, J.A. and S. Garcia. Observed patterns in multispecies fisheries. In: R. May, ed. Exploitation of Marine Communities: Report of the Dahlem Workshop. 1-6 April 1984, Berlin. Springer Verlag Life Sciences Research Report 1984; 32:155- 190.

41. Sherman, K. Can Large Marine Ecosystems be managed for optimum yield? In: K. Sherman and L. Alexander, eds. Variability and Management of Large Marine Ecosystems. AAAS Selected Symposium 1986; 99: 263-267.

42. McNeely, J.A. Common property resource management or government ownership: Improving the conservation of biological resources. Intl. Rel. 1991; 10(3):211-225.

43. Gjerde, K. and D. Ong. Marine and coastal biodiversity conservation. A report for the Worldwide Fund for Nature, Gland, Switzerland, 1995.

44. Mann, K.H. General concepts of population dynamics and food links. In: O. Kinne, ed. Marine Ecology IV: Dynamics. Chichester, England, Wiley Interscience, 1978; 617-704.

45. World Resources Institute (WRI). The State of the World. Washington, DC, World Resources Institute, 1996.

46. Hoagland, P., Y. Kaoru and J.M. Broadus. A methodological review of net benefit evaluation for marine reserves. World Bank Environment Dept., Pollution and Environmental Economics Division. Env. Econ. Ser. 1996; 26.

47. Hardin, G. Paramount positions in ecological economics. In: Ecological Economics: The Science of Sustainability. R. Costanza, ed. Environmental Economics, 1991; 47-57.

# CHAPTER 2

# THREATS TO OCEANS AND COASTAL AREAS

## SECTION 1. HUMAN IMPACTS ON MARINE AND COASTAL SYSTEMS

We still tend to think of ocean systems as having limitless resources, with an infinite capacity for resilience in the face of environmental pressures and change. The sea's vastness and its great mysteries–mysteries that exist because our understanding of marine ecology is decades behind that of land–obscures the fact that we are everywhere depleting resources and degrading systems in ways that are harmful and sometimes irreversible.[1] The degradation we are witnessing has occurred both as a result of direct impacts, as in the case of overfishing and conversion of coastal habitat, and of indirect impacts such as that caused by terrigenous and riverborne pollution.

Using living marine resources and ocean space does far more than merely deplete the stock of the target resource or momentarily preclude other uses of that space. Impacts from resource use range from the obvious, direct effects of exploitation such as diminished stocks and possible overexploitation, to other more indirect impacts related to how much of the resource is being extracted. Such effects include reduced stocks of organisms caught as incidentally or as by-catch, decreased availability of food for non-target marine organisms and, in the extreme, food web imbalances such as biomass flips.[2,3] Such imbalances can occur even when fishing pressure is relatively low: when key components of the marine ecosystem are removed, as is the case with naturally rare top predators that regulate food web dynamics below them, reverberations occur throughout the whole system. These cascading effects are poorly understood and difficult to control when sectoral management focuses attention solely on the target species.

Harvest also impacts marine systems not only according to how much resource is removed but how it is removed. Such negative impacts

*Marine Protected Areas and Ocean Conservation*, by Tundi Spring Agardy.
© 1997 R.G. Landes Company.

include loading the system with nutrients from fish-processing or wasted by-catch, changes to benthic or coastal habitats, loss or degradation of reproduction or nursery areas, disruptions of nutrient cycling and lowered water quality. Use of ocean space for non-extractive industries can have similar indirect effects. And in terms of magnitude of cumulative impacts over time, land uses that increase pollution of nearshore areas through degradation of river systems, sedimentation of seawater from deforestation or other forms of run-off acceleration, introduction of nutrients through agriculture and sewage, changes in hydrology due to water diversion or damming, as well as introduction of toxic pollutants, are responsible for much of the degradation of coastal systems we are witnessing today.

It is clear from the many signs of ecological impairment and biodiversity loss already apparent that we have altered the structure and function of coastal systems in many parts of the world. Most of the world's major fisheries are now overfished, many fisheries stocks have been extirpated and some heavily utilized areas are thought to be on the verge of ecological collapse. But not only are we chronically impacting our valuable and irreplaceable ocean areas, we seem to be driving the stakes straight through the heart of marine ecosystems. Most deleterious impacts occur in nearshore areas, where the bulk of areas critical to ecosystem functioning and productivity are found. Nutrient loading and mixing is highly concentrated in the nearshore areas, and most of the marine food web so critical to human population is based on this coastal productivity.[4] The great majority of fish, mollusc and crustacean breeding areas occur in estuarine or other coastal habitats, and the same is true for seabirds, shorebirds, sea turtles, marine mammals and other marine vertebrates. Sadly, these are the areas that suffer most from overexploitation and the indirect degradation caused by human activity on land and at sea.

In large part, the inability to use marine resources sustainably and control indirect degradation of coastal areas has to do with a worldwide perception that the oceans are a global commons. Although a sense of responsibility for impacts and stewardship for coastal areas in some human communities is strong, competition for ocean space and limited resources undermines this basis for rational use. Fully two-thirds of the world's population currently lives in the coastal belts of continents, and that figure is rising steadily (Figure 2.1).[5] More and more people are depending on marine sources of protein, and more and more national economies are dependent on coastal and marine-based industry. When a collective perception emerges that a resource or set of resources is both free for the taking and under the threat of exploitation by others, previously sustainable patterns of use can degrade into an anarchic system of misuse and overexploitation. This potential for destruction becomes accentuated when bounds of governmental jurisdiction over marine resources and space are in question, as they are in most parts of the world today.

There are two broad categories of threat to marine ecosystems and biological diversity: so-called "pressure factors" and "enabling factors."[6] Pressure factors include exploitation of resources, growing use of ocean and coastal space, human population growth and poverty. Enabling factors include trade policies, the undervaluing of national capital in regard to marine resources and inequitable access, benefit-sharing or burdens of cost of environmental protection. These factors, of course, contribute equally to terrestrial environmental degradation and loss of biodiversity, however, the nature of the problems in the marine realm more complex and the magnitude of the problems is often exacerbated by our relative inexperience in conserving marine ecosystems.

## SECTION 2. DIRECT THREATS TO MARINE SYSTEMS

### 2.1. OVEREXPLOITATION OF MARINE RESOURCES

The oceans and coasts harbor extremely complex ecosystems, with geographically widespread linkages and hierarchically nested spatial and temporal scales.[7] Such scales occur in terrestrial ecosystems as well, but the fluid nature of the marine environment and its exaggerated connectivity makes investigation of how communities are organized in space and time imperative to our understanding of ocean functioning. The important ecological processes that create the conditions necessary for the maintenance of the ocean's great biodiversity and immense productivity are closely linked across these scales.[8] Despite this, we tend to want to model marine systems as if they were fully deterministic and conveniently simplistic, ignoring the realities of connectivity, complicated recruitment and population dynamics, and inherent uncertainty.[9,10] As part of this desire to want to keep things simple, we cling to the mistaken idea that the acceptability of resource extraction or use can be evaluated by looking at the resource in question while ignoring all other parts of the system.

Our capacity to find and harvest living marine resources has now exceeded the natural capacity of coastal and open ocean ecosystems to replenish them. The first sign of such overharvest is commercial extinction–and if harvest continues at prior rates, this can lead to true species or population extinction. However, extracting living resources from the sea causes wider ranging impacts that reverberate through entire marine ecosystems. Indirect impacts of exploitative activity include the loading of the system with discards, including fish-processing waste and by-catch, habitat alteration, loss or degradation of nursery areas, possible impacts on nutrient cycling through nutrient loading, and lowered water quality in terms of chronic pollution by fisheries operations or single pulse pollution in catastrophic accidents.[11] As mentioned previously, the cumulative stresses caused by these and other impacts in tandem can cause long-term genetic and life history alterations in target and non-target species that influence marine productivity and ecological services.

*Fig. 2.1. Highly developed coastline, Monaco. Photo by T. Agardy.*

Though we commonly think of fish and fish products when we think of marine over-exploitation, these are not the only kinds of resources to suffer from overuse. Mangrove wood and marshland peat, for instance, are often harvested at rates beyond their intrinsic regenerative capacities. Other non-living resources are renewable but only in extremely long time frames, such as limestone formed from coral reef organisms. The use of resources such as these can quickly move from low-intensity, low-impact use to wholly unsustainable use that results in reverberating impacts throughout the ecosystem.

Rather than present a review of the literature concerning the range of negative impacts that resource exploitation and overexploitation has on ecosystem functioning, the following discussion attempts to dispel some common myths and misconceptions about human impacts on the seas.

### Myth #1. The main effect resource extraction has on marine life is to deplete the target stock.

What few good time series data have been collected on the wider ecosystem impacts of fisheries and other living marine resource exploitation suggest that secondary effects are extensive and potentially severely detrimental to the maintenance of biological diversity. Despite the fact that pelagic food webs are well-dampened by broad diets and are, relative to terrestrially based food webs, somewhat unstructured, removal of certain segments of the food chain can have ramifications throughout the community.[12,13] Northridge (1991) reviews studies of European fisheries and summarizes the impacts exerted via: 1) the removal of prey species–reducing food availability to predators; 2) the removal of predators–causing population changes in those species controlled by predator abundance (e.g., Goeden, 1982); and 3) disruption of equilibrial systems–increasing post-disturbance recovery times and potentially reducing overall productivity.[14,15]

Removal of prey has impacts on those predator populations that are food-limited or that are behaviorally or physiologically affected by changes in food distribution.[16] A popular example of the far-reaching nature of such effects is that of the decline of Peruvian seabirds following decimation of anchovy stocks offshore. This example, though it has found its way into the popular scientific writing, is confounded by shifts in the El Niño Southern Oscillation. A better example may be the reduction in abalone availability and its impact on sea otter populations in California.[17] Investigating the magnitude of such effects is complicated, since it is not only food availability in terms of quantity but food availability in terms of access that acts as a forcing function for many predator populations. For example, tightly schooled land lance and krill are important to large cetaceans, which are not able to collect up widely dispersed prey. Other examples are extensively reviewed by Dayton et al (1995) and Northridge (1991).[11,14]

Removal of predators also has cascading effects. Target predator species tend to be ones that have important ecological roles in the community.[18] The functional removal of great whales in the Southern Ocean and sea otters in kelp forests has had cascading effects.[17,19] Another noteworthy example is the depressed diversity of intertidal organisms that resulted from the experimental removal of keystone starfish species in Washington State (USA).[20] In Chile, massive intertidal changes have been shown to result from removal of predators by fishermen.[21] For coral reef systems, removal of carnivores (piscivorous fishes, for example) can lead to decreased overall fish diversity, algal overgrowth of corals and release of previously controlled populations of sponges and tunicates.[22] However, benthic systems may be more responsive to the removal of predators than previously thought, as illustrated by the possible impact that Alaskan King Crab harvest has had on other members of the benthic community. Some studies have demonstrated significant differences between coastal and distant water rock reef systems that they attribute to fisheries-induced removal of predators that then acted to release a different suite of benthic predators such as echinoderms and crabs.[11]

**Myth #2. Modern technology is so efficient, particularly in marine fisheries, that incidental catch and secondary impacts are virtually non-existent.**

Modern commercial fishing technology is indeed a dark wonder. Navigational systems, depth finders, recording fish-finding apparatuses and the like are now so sophisticated that target fish have virtually no place to hide. We have become the supreme hunters of the sea. This is not necessarily good, although the large profit margins exhibited by parts of the commercial fishing industry are defended by economists as demonstrating the economic efficiency of the industry. Modern commercial fishing methods, despite their ability to increasingly satisfy human needs for food, have the following two major negative impacts on marine ecosystems, independent of the effects described above having to do with how many fish are extracted: 1) ecological alterations in the way species are distributed; and 2) physical alterations to habitat.

As stated above, many marine predators, especially pelagics, depend on the availability of concentrated schools or balls of prey; when these patches are thinned in distribution, predator populations can decline or even become threatened.[23] However, fishing does not merely alter the population structure and distribution of target species. Even today with our technological marvels, an enormous amount of biomass is caught, killed and wasted as by-catch.[24] Some studies show that discard biomass often equals or vastly exceeds that of landings, as in the case of shrimp fisheries the world over. This by-catch problem recently entered the realm of public consciousness when the spotlight became focused on the incidental catch of whales (Hofman, 1990), dolphins (Norse, 1993), porpoises (Read and Gaskin, 1981), sea turtles (Agardy, 1992b) and some other high-profile organisms.[25-28] Yet few people recognize the widespread nature and overwhelming magnitude of the problem. There are indeed good biological reasons for additional, highly warranted concern when the affected species are sharks, rays, large groupers, or deep-sea organisms, all of which have life history characteristics of low fecundity, delayed reproduction and no parental care of young.[11]

Regarding habitat disturbance, many of the most modern fisheries utilize gears that have geographically widespread and long-lasting impact. Benthic trawling disturbs vast areas of benthic sediments and effectively obstructs the critical role that bio-perturbing benthic organisms play in system function.[29,14] Extensive changes related from six decades of intensive benthic fishing have been documented in the Wadden Sea off the coast of Germany,[30] while others have reported trawling-induced destruction of benthic communities in the English Channel.[31] Physical destruction of seagrass meadows and other critical areas has ramifications not only for the seagrass community but also for reef and pelagic organisms that use the habitat during some portion of their life histories, especially their recruitment biology. There are myriad types of trawls, dredges and traps that sit on or are dragged across the benthos.[32] As the technology grows in sophistication, so it grows in causing widespread negative impact, not the opposite as some might claim.

Some fishing practices are inherently destructive, even at low levels of use, such as occurs in artisanal and even subsistence fisheries.[33] For instance, use of dynamite and other explosives for harvesting coral reef and schooling fishes causes severe and often irreversible damage to habitats and the ecological communities they harbor. This phenomenon has spread from Southeast Asia, where it is said to have originated, to virtually every tropical sea. Non-selective poisons are also used to denude wide areas of nearshore oceans, causing lingering impacts as some toxins remain in the system and become assimilated and bioaccumulated up the food web. Poisons also cause major damage when they destroy reef-building corals that provide the physical framework for the reef community.

Even modern gear that is relatively efficient and non-destructive when properly used can have major negative impact when it is lost at sea. Gill nets, fish traps and even monofilament lines devoid of lures or bait continue to "ghost fish," and many non-target organisms also suffer high mortality rates from entanglement in such gear.

**Myth #3. Natural variability in marine population dynamics means that human impacts will never have more than a negligible effect.**

Many aspects of marine populations and ecological processes have an inherent variability; this underlies our difficulties in identifying and describing forcing functions for marine systems.[34] Patchy distribution of resources and uneven availability of suitable habitats for many species mean that survival, and thus the underlying population dynamics for most species, are based on chance events. Many species have only small natural windows of opportunity to feed or breed, and even small impacts can adversely affect species if this already narrow opportunity is compromised. For instance, larval recruitment of corals appears to be largely a "lottery" process.[35,36] As is the case above, some anti-environmentalists have tried to use this natural variability as an excuse not to manage or conserve. Unfortunately, it is an excuse we cannot afford to make for ourselves. Negative anthropogenic impacts can have catastrophic impacts on even widely variable systems, particularly when the impacts are exerted on populations that are close to threshold conditions for extirpation.[37] And cumulative impacts can occur when even low level detrimental effects become additive to increase the magnitude of the stress and its consequences, especially when each of these low level effects are repeated or chronic.

Ecologists have long admitted a kind of "physics envy:" their laboratory is a highly unpredictable place where experimental controls and replications are difficult and sometimes impossible to achieve.[38] However, this does not mean that ecological science is any less "hard" or useful. In the sloppy world in which we live and work, the signal to noise ratio is often so small as to make the signal barely discernable.[39] Nevertheless, we must strive to take note of signals to avert environmental disasters and guarantee our own survival. Science is still badly needed to support resource use and management decisions.[40] In doing so, we must distinguish between what is science in the service of management (which exists expressly to be monitored, questioned and evaluated) and what is blind, almost religious belief that we (however foolishly) might want to accept without question.[41]

**Myth #4. The inherent stochasticity of marine systems precludes management of these systems.**

All ecological systems are composed of non-linear relationships, especially marine systems with their more open patterns of dispersal.[42] Indeed, researchers are starting to discover that many marine systems exhibit so many alternate or multiple equilibria that even labeling them "equilibrial" may be a fallacy.[43,44] Some decision-makers have used this apparent stochasticity as ammunition against attempts to develop scientifically based management plans for resource use and habitat conservation projects. Recent papers in scientific journals (especially Ludwig et al. 1993) have increased the intensity of these debates.[45] The question now being asked is whether we as conservationists are deluding the public by claiming that we can manage marine systems (or any natural systems, for that matter) for sustainability.[46] First, it is important to remember that we never manage natural systems no matter how deterministic those systems may be, we manage ourselves and try to mitigate the negative impacts that humans have on the rest of nature.

**Myth #5. The oceans are so vast and living resources so abundant that man will never deplete them or adversely impact ecosystem function.**

It is clear from the many signs of degradation that are already visible that this is faulty thinking. Not only are we chronically degrading our valuable and irreplaceable oceans, we seem to driving the stakes straight through the hearts of marine systems. Most of the world's fisheries are now overfished; many fisheries stocks have been extirpated and some heavily fished communities are even thought to be on the verge of ecological collapse. The truly frightening fact about the impacts that humans have on oceans is that most of these impacts occur in the nearshore areas, where the bulk of the marine "vital organs" are located. Nutrient loading and mixing is concentrated in the neritic zone and upwelling areas, and most extensive marine food webs are based on this nearshore productivity.[47] The large majority of fish and crustacean breeding and nursery areas occur in estuarine or other coastal habitats; the same is true for shorebirds, seabirds, sea turtles and other marine vertebrates. Sadly, these are the areas that suffer most from overexploitation and indirect degradation caused by anthropogenic activity on land and at sea.

Overexploitation has caused the complete restructuring of marine ecosystems in some parts of the world. In the George's Bank area of the northwestern Atlantic (off the coasts of Massachusetts, USA and Nova Scotia, Canada), massive exploitation of demersal fishes in the cod and haddock family has shifted the most abundant components of the food web by biomass to the scavengers rather than the secondary consumers. More dramatically, massive overfishing in the semi-enclosed Black Sea (bordered by Ukraine, Russia, Georgia, Turkey, Bulgaria and Romania) has acted with other degrading forces to cause the extirpation of many fish populations and marine species. The Black Sea is now considered to be a prime example of a system transformed by ecosystem collapse wrought at the hands of man.

**Myth #6. Fishing and other marine resource use is naturally self-regulated.**

Long ago when fishermen were greatly fewer in number, habitats were fairly pristine, the need to use resources was nowhere near what it is today and fishing was largely a self-regulated enterprise. When one fish stock was depleted by harvest, fishermen naturally switched

to more abundant stocks. This allowed the depleted stock to recover
to pre-disturbance levels, with the typical result being that the impacts of exploitation were short-lived and inconsequential. The situation today is different because the numbers of users have increased,
pressures to compete and potentially overharvest are unavoidable, over-
capitalization of commercial fleets pushes operators to maximum economic returns even as the resource base is destroyed, and stewardship
among local users has virtually evaporated.[48] With the erosion of responsibility and forward-looking planning, and in the absence of clear
jurisdictions for marine resource management, whatever self-regulation
that may once have existed is now long gone. Bilateral and multilateral agreements are desperately needed to fill the void, but these
will only be embraced by the international community when governments
recognize the severity of the marine conservation problem.

**Myth #7. Aquaculture will save the day.**
   Aquaculture has come a long way towards meeting human demand
for fish products and increasing the stakes that governments have in
ocean systems. Unfortunately, most aquaculture operations produce luxury
items that command a high market value but are not sufficient in
meeting basic human needs. Furthermore, aquaculture is not a panacea.[49] Many culture operations put enormous pressures on coastal habitat,
often altering and sometimes even removing habitats from natural use.
Shrimp pond construction in parts of South America, for instance,
has destroyed much of the mangrove forest that previously supplied
wild shrimp stocks with important nursery habitat. Secondary impacts
of aquaculture also include increased pollution from normal operating
and accidents, changes in genetic and age structure of wild populations when cultured or captive-bred animals are released back into the
wild and increased incidence of disease caused by aquaculture-related
spread of pathogens. An even greater problem may be that decision-
makers sometimes have even less incentive to conserve natural habitat
if they think that mariculture operations will meet their subsistence or
economic needs. Figure 2.2, for instance, shows the mariculture of
highly valuable giant clams that is taking place throughout the South
Pacific, possibly removing incentive to protect wild stocks of the species and the natural habitats that support them.

## 2.2. HABITAT MODIFICATION THROUGH INHERENTLY DESTRUCTIVE PRACTICES

   Certain uses of marine resources and ocean space are by their nature inherently destructive to marine biodiversity and the ecological
processes that maintain diversity. Perhaps the most dramatic example
of intrinsically harmful exploitation of marine resources is the harvest
of coral reef fishes using explosives. A single blast can decimate thousands

Fig. 2.2. Giant clam aquaculture operation in the South Pacific. Photo by J. Dewey.

of years of cumulative coral reef development and can preclude any regeneration of reefs in the vicinity of the explosion. Other destructive practices include mining live coral for construction materials (or live rock for the aquarium trade), use of poison for fishing, bottom trawling, deep seabed mining, nearshore sand mining, mangrove deforestation and disposal of hazardous waste in certain conditions.

Inherently destructive fishing and development practices have impacts that range from localized and relatively small to regional and long-lasting. The use of dynamite and other explosives to harvest coral reef fishes for food and for the growing aquarium trade can cause wide areas of destruction that effectively decimate thousands of years of coral reef growth. Although the effect on stony corals themselves is the most apparent sign of destruction, the ecological impact far exceeds the loss of coral colonies themselves. The destruction of the physical framework of the reef carries with it the attendant loss of symbiotic and mutualistic relationships between key species in the reef community, and the resulting ecological collapse can extend far beyond the blast site itself. Furthermore, destruction of the reef structure can alter current patterns and lead to shoreline erosion and instability, so that the impact extends beyond the shallow waters to the coast itself. If changes to oceanographic processes are extensive, regeneration of the reef and its biota may never occur and the area may remain a low productivity wasteland indefinitely.

The use of poisons to harvest fish for food and for the aquarium trade similarly has effects that far exceed what is readily apparent.[33] In the tropical Pacific region, for instance, use of cyanide for exploitation of reef fishes not only kills and stuns the target species but also inhibits growth of corals, sponges and epifauna on the reef. This in turn manifests itself as lowered productivity and, in extreme cases, in ecological collapse of an entire fringing or patch reef community. Some poisons have an exceedingly long residual time in sediments and can cause chronic impacts over many years.

In terms of spatial extent of destruction, however, large scale commercial bottom trawling for benthic fishes far exceeds impacts caused by dynamiting or poisoning. The dragging of heavy gear such as trawl doors over soft seafloor substrates destroys the sea life on the ocean floor and disrupts ecological processes that are critical to maintaining marine productivity and diversity.[11] For instance, marine worms that inhabit the benthic muds play a critical role in turning over and filtering sediments, a process known as bioperturbation. When such worms are killed by trawl gear, sediments become rock-like, inhibiting the colonization of the substrate by epifauina and disrupting the exchange between sea floor and water column. Fish eggs and larva that live in the benthic areas for a portion of their life histories can be effected by the environmental changes, such that the end result is lowered fisheries production in what was previously a very fertile fishing ground.

Sediments that are released into the water column as fishing gear is dragged over the bottom also cause problems in areas beyond the drag site. For example, sediments can suffocate filter feeding organisms such as anemones, corals, tube worms, etc., thereby removing important components of the food web from the ecosystem. There is recent evidence to suggest that such trawling effects are extremely long lasting, and in some cases cause permanent damage to the ecosystem.[49]

By far the most insidious, chronic direct impact that humans have exerted on the marine ecosystem has been the cumulative alteration of coastal areas.[50,51] Such alteration ranges from the small scale and seemingly innocuous, such as the construction of breakwaters, jetties and sea walls to the large scale and dramatic examples of habitat loss through conversion of coastal areas into urban and industrial landscapes. In between these two extremes are the conversion of mangrove forest to agricultural land, infilling of marshland for construction of homes, deforestation and peat bog mining and similar activities that alter the nature of the ecological communities and alter the functioning of the system. Nearly half of the world's salt marshes and mangrove swamps have been cleared or drained for coastal development, while some 10% of the world's coral reefs have been eliminated. Estimates of worldwide beach erosion from anthropogenic activities hover around a staggering 70%.

Coastal development impacts marine ecosystems and their intrinsic biodiversity in direct and indirect ways. In the simplest cases, construction of housing, infrastructure, or industry displaces species and converts natural coastal environments into artificial areas. In the United States and other developed countries, coastal development has typically involved the infilling of salt marshes and other coastal wetlands, resulting in loss of species and areas critical to larger ecosystem function. For instance, conversion of southern Florida's wetlands to areas for housing developments has resulted in the local extirpation of hundreds of species, loss of stopover and feeding areas for migratory birds and disruption of ecological services such as maintenance of hydrological balances, filtration of water and nutrient loading to the nearshore ecosystem.

Coastal development impacts extend well beyond the footprints of the construction or infilling area itself. Nearshore hydrodynamics, so critically important to maintaining diversity and productivity of coastal systems, can be dramatically altered by coastal engineering projects and habitat conversion. These impacts are occurring at alarming rates throughout the world. Again, what is especially frightening about the scale and rate of habitat loss is that the most vital, ecologically important areas are the ones that are being hit the hardest by direct and indirect impacts of coastal development. As mentioned previously, it is as if we are targeting the vital organs of the earth ecosystem in wreaking havoc and destruction in the coastal zone.

Coastal habitat alteration and loss will only increase in years to come, as demands for space and resources increase. Throughout the globe, human population growth rates are increasing fastest in the coastal belt, and migrations to coastal areas compound the population problem. When inland regions become overdeveloped and agricultural lands become overused, further migrations to coastal areas take place. In many parts of the world, coastal industries have become "employers of last resort" for people who cannot earn a living or find subsistence in interior regions.[52] As populations in the coastal zone continue to rise, demands for space, food, energy and employment will put increasing pressure on coastal resources. In most cases, such demands will not be met because coastal ecosystems will already be compromised by degradation.

A note of caution should be sounded in the population versus coastal pressure debate, however. In some cases, there is an inverse correlation between population density or the number of people living on the coast and the condition of the coastal environment. When damaging industries are sited in coastal areas, release of pollutants and toxins can cause far greater damage to the marine ecosystem than urban development or excessive harvest. In assessing threats to the marine biological diversity, therefore, it is advisable to refrain from simplistic indices such as population density and instead look at magnitude of destructive impacts themselves.

## 2.3. TOXIC DISCHARGES (POINT SOURCE POLLUTION) AND OCEAN DUMPING

Although several international instruments exist to regulate the discharge of hazardous materials at sea (e.g., London Dumping Convention, MARPOL, regional agreements), such materials continue to find their way into marine systems through illegal activity, national policies that allow disposal of certain materials within the exclusive economic zones (EEZs) of that country and through accidents at sea.[53] Such hazardous waste includes *inter alia*, radioactive material, persistent toxics such as organochlorides, hydrocarbons, and sewage sludge. Sewage sludge introduces pathogens and persistent toxins into the ocean environment, and at concentrations as low as 0.1% causes mortality to herring and cod eggs. Plastic materials and other debris are sometimes deliberately discharged as well, causing massive mortality to marine life as organisms ingest it or become entangled in it. Dredge spoils from maintenance of navigation channels and coastal development practices can cause the resuspension of toxics into the water column. While over 80% of material dumped at sea is dredge spoil material, at least 10% is estimated to be contaminated from urban, industrial and shipping-related toxic chemicals.[51]

A particularly alarming facet of chemical pollution of the seas is the recent evidence that land-based industries have led to the release of thousands of types of toxic chemicals into riverine and coastal environments. Some of these chemicals have been found to be endocrine disruptors–compounds that mimic the effect of hormones in the bodies of humans and wildlife alike.[49] Such toxics accumulate in organisms that occur near the bottom of food webs, and can kill or otherwise impair the organisms (including humans) that feed on contaminated fish, molluscs, or crustaceans. The presence of these ubiquitous toxic compounds is only now beginning to be documented in coastal areas worldwide. It will be years before we come to understand the full ecological and human health impacts that they cause.

## SECTION 3. INDIRECT THREATS TO COASTAL AND MARINE SYSTEMS

### 3.1 POLLUTION FROM LAND-BASED SOURCES

As is the case with overexploitation of marine resources, habitat alteration can be the result of a direct conversion process (e.g., urbanization and port construction), or an indirect effect of human activity in either the coastal zone itself or in areas linked to that coastal zone. For instance, siltation of rivers caused by poor land use practices (clearcutting of timber, for instance) can cause dramatic conversion of nearshore habitats, and an attendant loss in biodiversity that is dramatic in scale.

Not only are these sorts of impacts driving the bulk of marine environmental degradation and loss of biodiversity values, these are the kinds of impacts that are the most difficult to assess and regulate.

There are myriad ways that humans indirectly impact marine biological diversity and the ecosystems that sustain it. Pollutant release, run-off of toxics, excessive fertilization of nearshore waters from agricultural runoff, release of alien species and changes in hydrology in river systems all cause dramatic changes to the ecology of nearshore and, in some cases, offshore ecosystems. A typology of indirect threats is given in Table 2.1; however, the magnitude of impact from any one of these kinds of threats is often difficult, if not impossible to determine.

The situation with respect to indirect threats is also compounded by the fact that most pollution events are not single isolated episodes, but occur over time. Some of the most insidious threats to marine biological diversity are those that are chronic, but often so small in magnitude at any single point in time that they escape detection. Additionally, most of these threats occur together and compound one another. Such cumulative impacts are notoriously difficult to tease apart and identify true cause and effect.

*Table 2.1. Types of marine pollutants and sources*

| Type | Primary Source | Other sources |
| --- | --- | --- |
| Nutrients (phosphorus- and nitrogen-based compounds) | sewage | land use |
| Sediments | dredging | land use |
| Pathogens | sewage | livestock waste |
| Persistent toxins (heavy metals, PCBs, DDT and derivatives) | industrial and urban wastes | seepage from landfills, etc. |
| Hydrocarbons (oil) | automobiles | shipping, accidents at sea, seepage |
| Plastics | land-based sources, fishing operations | cargo ships and industry |
| Radioactive wastes | industrial and military wastes | atmospheric fallout |
| Thermal pollution | power plant cooling systems | industry |
| Noise pollution | supertankers and other vessels | machinery |
| Introduced species (aliens) | ballast water projects | canals and fishery |

## 3.2. INTRODUCTION OF ALIEN OR EXOTIC SPECIES

The transport and ultimate release of species not indigenous to an area is becoming an increasingly serious threat to the world's marine and aquatic systems. Exotic or alien species are often superior competitors that displace local fauna and do so extremely quickly (as typical of colonizing species). Alien species also cause "genetic pollution," by disrupting the genetic diversity of natural populations of organisms and altering the structure of species assemblages. This unnatural stirring of the world's gene pools is accelerated by technology: as ships become bigger and faster their ability to carry alien species in ballast water and the rate of their dissemination is increased. With a world shipping fleet of 35,000 ships, several thousand different ballast water-borne species may be transported to new habitats each day.

There are many examples of the ecological catastrophes wreaked by such alien species. San Francisco Bay in the western United States has been receiving alien marine species at the rate of one every fourteen days; many of these organisms survive, settle and eventually displace other native species.[54] Another example can be found in northern Africa, where the opening of the Suez Canal led to the spread of more than 250 species from the Red Sea to the Mediterranean. Some of these introduced species, like the Red Sea jellyfish, have displaced local and commercially important species, depressed fish catches, clogged intake pipes of coastal power plants and impacted tourism at Mediterranean beaches. In the Black Sea, ballast-borne introduction of the ctenophore *Cnemiopsis* from the eastern U.S. has caused the decline and population extirpation of at least 28 commercially important species of fish, and has contributed to the "desertification" of a once productive and economically important sea.

## 3.3 FRESHWATER DIVERSION IN WATERSHEDS

The damming of rivers for purposes of irrigation or energy generation can cause the salinity of downstream estuaries to change dramatically. Biodiversity is impacted as a consequence, as only salt-tolerant species are able to continue to utilize the environment. For example, the large scale reduction in biodiversity and overall degradation that has occurred in the last few decades in Florida Bay (USA) has been directly attributed to the reduction in water flows through the Everglades wetlands area and corresponding loss of freshwater input into the bay (National Oceanic and Atmospheric Administration, 1995).[55] The impact to the ecology of the coastal area, however, is not confined to the deltaic or estuarine regions. Often, highly productive and important estuaries that act as nursery grounds for fisheries become hypersaline and can no longer support fisheries production. Productivity then declines in a wide region, underscoring the ecological connectivity between various habitats within a coastal ecosystem. The extent and magnitude of this threat worldwide has not been assessed, however, such large scale civil engineering projects have taken place in virtually all coastal countries with large rivers and ecologically important estuaries. Noteworthy examples of hydrological civil works far upstream affecting coastal fisheries production and biological diversity include the Everglades water diversion projects undertaken by the U.S. Corps of Engineers, the construction and operation of Mahaweli dam in Sri Lanka and several irrigation and hydroelectric programs in North and sub-Sarahan Africa.

## 3.4. GLOBAL CHANGE PHENOMENA: SEA LEVEL RISE, WARMING, CHANGES IN PHYSICAL OCEANOGRAPHY

Conservationists have long been worried that low level yet chronic anthropogenic tinkering with the biosphere, such as loading the atmosphere with ever-increasing carbon dioxide emissions, may undermine the abilities of ecosystems to function properly. When these impacts occur simultaneously with other global phenomena, like ozone depletion, the compounding effects can be disastrous. The wearing down, as it were, of the ecosystems' "health" (this is in quotes because we really don't yet know what constitutes a healthy ecosystem) means that otherwise resilient systems may not be able to cope with the natural changes that are a part of the dynamic, ever-evolving living world.[1] We feel confident that ecosystems have an intrinsic threshold for tolerating insidious degradation, and that once this threshold is surpassed, systems become wobbly, headed towards demise and collapse.

This may seem elementary and intuitively obvious. Yet, there is to date little hard evidence showing how ecosystems respond to human-induced change; in fact, our knowledge about how components of any system interact is disgracefully poor. We retain a bias for a structural, rather than functional, view of the world. Scientists still monitor the structure of ecological communities, e.g., numbers of organisms, height and percent cover of plants, etc. to indirectly ascertain the ecosystem's condition. This is an insufficient proxy at best, and conservation plans still focus on maintaining structure, rather than safeguarding critical processes.

New international attention that has been given to the global warming debate, and the promise of financial support to help address some unanswered questions about it, may help to remedy the paucity of knowledge about ecosystem and biosphere functioning. Even if you disagree with the worst-case scenario predictions on the greenhouse effect, tackling the question of "What if?" is a useful exercise. And though this question is currently being directed at highly specialized groups of ecologists who work in relatively narrow biotic domains, one can envision a time in the not-so-distant future when these specialists will come together to talk about inter-ecosystem interactive impacts.

The prospective impacts of global climate change that scientists are currently addressing include sea level rise (or relative sea level change), increased $CO_2$ levels, increased temperatures, increased ultraviolet (UV) radiation, changes in sediment dynamics (and the relation between hydrological/sediment flux and sea level rise), increased primary production and nutrient loading, increased storm frequency and intensity and changes in nearshore salinity. Secondary impacts from global change on oceanic systems include loss or displacement of important fisheries production and increased risk to human welfare as catastrophic events become even more difficult to anticipate.

Some preliminary conclusions can be drawn from the relatively new field of global climate change investigation, including the following: 1) coastal mosaics are so complex that, for the time being, site-specific investigations are necessary to assess ecosystem vulnerability; 2) the net effect of global warming is negative on all systems investigated, particularly when warming impacts are compounded by other environmental degradation; 3) temperature changes, increased storm events and sediment changes will have the most negative impact on marine systems; 4) in nearshore systems, especially in the tropics, altered hydrology regimes (nutrient loading, sedimentation, pollutants and salinity changes) that result from global climate change may cause severe ecological damage; 5) our preoccupation with relatively well-studied, close nearshore habitats may distract us from looking at the full range of impacts predicted global climate change will have on the marine system and the earth as a whole; and 6) researchers should focus on metabolic responses at the organismal and community level and work towards a better understanding of systems ecology.[56]

Recent climate change studies have begun to shed light on the impacts we can expect on marine biological diversity and productivity. Direct impacts of global warming include loss of temperature-sensitive organisms, as occurred with the disappearance of gorgonian (soft coral) species off Panama following an acute warming event.[57] However, indirect impacts from warming are expected to be much more drastic.[58] These include disruptions of fisheries productivity and distribution caused by warming-induced changes in current patterns, changes in hydrology and chemical oceanography due to melting in polar regions and sea level rise.[57] The latter is particularly threatening to small island nations for two reasons: 1) such countries have a strong dependence on intact marine ecosystems for economic growth and subsistence; and 2) residents of such areas have nowhere to flee as coastal areas subside.

Future climate change will have potentially significant negative impacts on flagship marine species such as sea turtles and cetaceans and may hinder their population recovery.[59] The climate change effects can be classified into two sets of phenomena: global warming and ozone depletion. The former include warming-related changes to physical oceanography and attendant impacts on the distribution and abundance of food for such marine species, sea level rise impacts in the location of critical areas for endangered cetaceans and sea turtles, increased coastal precipitation affecting water quality of nearshore habitat and possible environmental changes affecting communication and social structure (especially important to highly evolved and social species). The effects of ozone depletion include similar changes in the food availability, changes in food quality and increased incidence of UV-related disease.[59] Much of this is speculative and awaits verification by interdisciplinary research by climate modelers, physical oceanographers and cetacean biologists. What is known with certainty is that negative impacts of climate change compound what is an already dire situation regarding whales, small cetaceans and sea turtles, and may be the gravest of all the cumulative threats affecting these endangered species and threatened habitats.[56] The implications for marine conservation warrant serious attention; future debate will have to focus on whether harvesting such species is sustainable given these new threats, what additional protection measures are necessary to counter these threats and how to optimally site future protected areas for species of special concern and vulnerable habitat types in a changing global marine environment.

## SECTION 4. MANAGEMENT CONSTRAINTS AND SOCIETAL ATTITUDES

### 4.1. INSUFFICIENT KNOWLEDGE/SCIENTIFIC UNCERTAINTY

The principle challenge confronting decision-makers wishing to conserve their marine biological diversity and thus keep options alive for deriving values from the coastal and marine environment is to develop management (despite uncertainties) that allows for optimal resource use while maintaining ecosystem integrity. All of this conservation activity should be undertaken with the ultimate goal of equitably distributing benefits of marine biodiversity conservation in society. But given the uncertainties that exist in our understanding of how marine systems are structured and how they function (much less the knowledge base that we have that would allow us to accurately predict what happens to marine ecosystems when we exert development and other pressures on them), can we realistically expect good resource management decisions arising from science-based management? The answer is yes, as long as intrinsic differences between terrestrial and marine systems are acknowledged and inherent uncertainties are accepted as part of the status quo.

There are fundamental differences between marine ecosystems and the better-studied and understood terrestrial analogues, as was discussed in section 1 of this chapter. The important features of marine systems that contribute to this dichotomy include: 1) marine systems are relatively poorly understood, having a much shorter history of study and presenting logistic difficulties for researchers; 2) marine and coastal systems have nebulous boundaries, both in terms of the outer bounds of systems and in terms of limits to ecological communities and populations of organisms; 3) marine systems exhibit enormous geographical and spatial scales, yet relatively fine temporal scales, meaning that although processes may be connected over wide distances, environmentally induced changes are common and almost immediate; 4) marine systems are driven by largely changeable and unpredictable physical processes; 5) marine systems have largely unstructured food webs, so impacts of resource removal on food chain dynamics is not easy to predict; 6) marine systems are characterized by varying degrees of linkage between communities of organisms in the water column and those of the benthos; and 7) marine system dynamics are generally more non-linear and stochastic than terrestrial systems, making generalizations difficult. These inherent characteristics of marine systems contribute to our failure to fully understand marine ecosystem function. This in turn implies that management decisions must be made in the face of enormous uncertainties and gaps in knowledge.[60,10]

Living with uncertainty is a fact of life. But our paucity of knowledge about marine systems can also be redressed by supplementing scientific knowledge with traditional knowledge that exists in coastal societies throughout the world. The information gained from western scientific study perfectly complements, and is complemented by, the knowledge developed by and passed on through generations of local people who rely on the seas.[61] Such knowledge is truly invaluable for decision-makers in need of information to help prioritize action and design effective plans for sustainable use and conservation. However, tapping that knowledge will require sensitivity—sensitivity on the part of sociologists to be fair in accessing that information and sensitivity on the part of conventional scientists to incorporate that information as fully as possible into their more 'rigorously' derived knowledge base.[62]

Few people would argue for methods of resource use that are inherently over-depleting, destructive or otherwise unsustainable. As discussed in the next chapter, figuring out what levels of use are truly ecologically sustainable is a difficult task, and requires that we rigorously define our objectives before attempting to derive what those limits might be. Scientific research is beginning to point us in the direction of sustainability, minimizing the potentially destructive impacts that fulfilling our resource needs has on the balance of nature.[63] However, there will always be scientific debate about values, methods and priorities.[60] We might well be suspicious if such debate were absent, because scientific consensus is largely a fiction.

Yet we still know very little. In the face of this scientific uncertainty, and in light of our impatience with incomplete knowledge, we would be wise to adopt a precautionary approach to err on the side of conservation, at least until we gain information we need to justify greater levels or more destructive kinds of use. While working to limit our impacts on ecosystem functioning, we must strive to gain better scientific information and make it accessible to decision-makers and managers. The usefulness of such information will largely reflect the quality of the scientific questions we ask and the kind of institutional, political, and financial support we give our researchers. Given that the pace of data accumulation and analysis is slow, some studies should be oriented towards answering management questions and fine tuning research so that basic research becomes applied research. Fuentes (1993) suggests, for instance, that we use science to set boundary conditions for the sustainability envelope, concentrating our efforts to determine what activities can clearly be labelled as non-sustainable.[64] Clear questions are needed about how we undermine ecosystem functions, how we can continue to use living resources sustainably, and how we can modify our behavior to ensure our survival and with it the rest of the planet. Also, we must set out to derive those answers as quickly as possible.

Thus, coastal and marine ecosystems present unique challenges to resource managers, conservationists and others wishing to exert some control on how humans and anthropogenic activity impacts the oceans. However, the inherent scientific uncertainties should not get in the way of progress in our efforts to conserve the marine environment, through marine protected areas or otherwise. What we are left with as a consequence of the nature of marine systems and their uncertainties is the need to: 1) create multiple alternate hypotheses in scientific investigation; 2) allow for some subjective interpretations and support common sense approaches (Johannes, 1989); 3) develop management approaches that can be adapted as necessary, and support management actions that are informative (so that we may learn from management, as well as from basic scientific research); 4) support management decisions that are reversible, and invoke the precautionary principle whenever possible; and 5) undertake and make available in an accessible form, information derived from cross-disciplinary scientific research.[65] Most importantly, while being honest with decision-makers and the public about how much we truly know about ocean ecology and the consequences of any proposed actions, we must move away from using uncertainty as an excuse for inaction.

## 4.2. FRAGMENTED MANAGEMENT OF COASTAL AND OCEANIC AREAS AND THEIR RESOURCES

The management of coastal and marine areas the world over is characterized by a patchwork of agencies with overlapping or redundant jurisdictions and little communication or coordination between vested interests. Rather than work effectively to manage ocean space, competing interests have led to a Balkanization of administrative bodies.[66] The management of endangered sea turtle species in United States waters is a good example: while in the 200 mile Exclusive Economic Zone (EEZ) they are protected by the National Marine Fisheries Service (under the National Oceanic and Atmospheric Administration), while nesting on shore they are protected by the Fish and Wildlife Service (Department of Interior) and while on the high seas they are effectively ignored. The case is even worse for habitat protection, which can fall under the auspices of any number of environmental protection, fisheries, energy, transport or military agencies.

Most coastal nations around the world, whether developing or industrial, exhibit a fragmentation in government when it comes to assessing the state of the marine environment or protecting its ecological integrity. This fragmentation is often the result of a sectoral approach to management, such that fisheries management is the domain of one agency or ministry, while coastal development for tourism is the domain of another and industrial development is the domain of yet an-other.[67] Security interests are most often segregated from development interests, which lead to a situation where a country's navy is pitted against other government agencies with their own marine jurisdictions.[68,69]

Admittedly, there are intrinsic problems in managing ocean envi-ronments. Firstly, it is impossible to draw direct parallels between the management of human-dominated and highly controllable landscapes (as occur in Europe, for instance) and management of the seas. In addition, the western style of resource management that provides one commonly used model of management relies heavily on command control legislation, whereby monitoring is focused on identifying violators.[70] Breaking the law is an integral component of command control and perpetrators choose whatever alternative is least likely to be enforced or where fines are the lowest. In marine regulation, command control is less feasible because enforcement and litigation are difficult, and non-point source effects are everywhere.

Currently the management of much of the world's marine envi-ronment is carried out at numerous levels of government simultaneously, including local, state, regional and national levels of government. At any given level various functions can be carried out through a wide array of agencies and organizations with little or no coordination among them. This means that not only is horizontal fragmentation a problem (i.e. federal overlapping jurisdictions as described in the sea turtle ex-ample above), but vertical fragmentation (i.e., lack of coordination be-tween local, state and federal governments, for instance) is also an issue. This fragmentation results in substantial inefficiency. This frag-mentation can also create conditions in which significant conservation and management issues receive inadequate attention.[71]

How can this problem of fragmentation and inappropriate com-mand/control legislation be rectified? One solution is to restructure government so that agencies have the ability to work together on is-sues of common interest. This is the philosophy behind Integrated Coastal Management (ICM) or Integrated Coastal Zone Management (ICZM), an idea that has become popular in the last decade. But marine protected areas can also help redress this problem by providing a test-ing ground in which to integrate management across all sectors and open up lines of communication and cooperation between previously sectoralized and isolated agencies.

## 4.3 GLOBAL COMMONS PERCEPTIONS

Part of our inability to limit our use of marine resources to levels that might be ecologically sustainable over the long-term has to do with the important perception that the oceans are a global commons.[72] Although the sense of stewardship in some (mostly lesser developed) coastal communities is strong and has been so for many generations, competition for coastal space undermines people's personal links to the sea.[73] It has been estimated that 70% of the world's population now lives in the coastal belts of the world's continents, and that figure is rising steadily. When a collective perception emerges that a resource or set of resources is both free for the taking and under threat from use by others, patterns of relatively sustainable use can erupt into an anarchic system of overexploitation and misuse.[74] This chaotic behavior can be amplified when the bounds of government jurisdiction over marine resources are in question, as they still are in many parts of the world.[75]

In coastal and marine areas, human activity is commonly unsustainable because common property perceptions lead to competition for resources and a "take what you can" philosophy.[73] Notions of stewardship, where they exist, are being undermined by the exponential growth of coastal populations and the subsequent lowering of coastal environmental quality. If, however, management can demonstrate to users its ability to preserve resources for sustained use, then sociological and political interest will embrace sustainable planning.[76]

### 4.4. SOCIETAL ATTITUDES

Societal attitudes towards the seas, lack of ownership and property rights, few precedents demonstrating effective management of ocean space and historically nebulous jurisdictions conferring rights and responsibilities for use of ocean resources and space all have contributed to poor coastal and ocean management in the past.[77] The situation can be summed up globally by the phrase "out of sight, out of mind," explaining in part why the oceans continued to be used indiscriminately as a sink for our wastes and a site for uncontrolled exploitation. The underlying attitudes also explain why marine conservation is decades behind terrestrial conservation, and why when the public and decision-makers think of the biodiversity crisis they overlook the seas and focus primarily on tropical forests.[78,71]

There are parts of the world, however, where historical use, cultural affiliation and societal attitudes do set the stage for future effective management of marine areas. In Oceania, for instance, marine tenure is an accepted part of traditional use of ocean space and resources, and responsibility for sustainable use of resources is already embedded in cultural norms. In the wider South Pacific, marine resources are so culturally and economically important that concern about the condition of the marine environment surpasses concerns about the state of the land and its resources.[48] In much of the industrialized world, customary law and an adherence to the U.N. Law of the Sea Treaty (UNCLOS) has codified a standard for jurisdiction and responsible use of ocean space, at least to the limits of the continental shelf (or the designated 200 mile EEZ).[79] Additionally, recent United Nations negotiations concerning Straddling Stocks, which grew out of the Rio Earth Summit (UN Conference on Sustainable Development held in

Rio de Janeiro, Brazil in 1991), show promise for stewardship of living resources that travel between jurisdictions of coastal countries or live in the global commons known as the high seas.

## REFERENCES

1. Barrett, C.W., G.M. Van Dyne and E.P. Odum. Stress ecology. Bioscience 1976; 26(3):192-194.
2. Sherman, K. Sustainability of resources in large marine ecosystems. In: K. Sherman, L. Alexander, and B. Gold, eds. Food Chains, Yields, Models, and Management of Large Marine Ecosystems. New York, Westview Press, 1991; 1-34.
3. Sherman, K., L.M. Alexander, and B. Gold, eds. Food Chains, Yields, Models, and Management of Large Marine Ecosystems. New York, Westview Press, 1990.
4. Smayda, T. Global epidemic of noxious phytoplankton blooms and food chain consequences in large ecosystems. In: K. Sherman, L. Alexander, and B. Gold, eds. Food Chains, Yields, Models and Management of Large Marine Ecosystems, 1991; 275 -308.
5. Ray, G.C. Sustainable use of the ocean. In: Changing the Global Environment. New York, Academic Press, 1988; 71-87.
6. World Bank. National Environmental Strategies: Learning from Experience. Washington, DC, World Bank Environment Department: Land, Water and Habitats Division, 1995.
7. O'Neill, R.V., D.L. DeAngelis, J.B. Waide, and T.F.H. Allen. A Hierarchical Concept of Ecosystems. Monographs in Population Biology 1986; 23. Princeton NJ, Princeton University Press.
8. Costanza, R., W.M. Kemp and W.R. Boynton. Predictability, scale, and biodiversity in coastal and estuarine ecosystems: implications for management. Ambio 1993; 22(2-3):88-96.
9. Mangel, M. Decision and Control in Uncertain Resource Systems. New York, Academic Press, 1985.
10. Mangel, M., L.M. Talbot, G.K. Meffe, M.T. Agardy, D.L. Alverson, J. Barlow, D.B. Botkin, G. Budowski, T. Clark, J. Cooke, R.H. Crozier, P.K. Dayton, D.L. Elder, C.W. Fowler, S. Funtowicz, J. Giske, R.J. Hofman, S.J. Holt, S.R. Kellert, L.A. Kimball, D. Ludwig, K. Magnusson, B.S. Malayang, C. Mann, E.A. Norse, S.P. Northridge, W.F. Perrin, C. Perrings, R.M. Peterman, G.B. Rabb, H.A. Regier, J.E. Reynolds III, K. Sherman, M.P. Sissenwine, T.D. Smith, A. Starfield, R.J. Taylor, M.F. Tillman, C. Toft, J.R. Twiss, Jr.,J. Wilen, and T.P. Young. Principles for the conservation of wild living resources. Ecol. Appl. 1996; 6(2):338-362.
11. Dayton, P., R. Hofman, S. Thrush, and T. Agardy. Environmental effects of fishing. Aquatic Conservation: Marine and Freshwater Ecosystems 1995; 5:205-232.
12. Dagg, M., C. Grimes, S. Lohrenz, B. McKee, R. Twilley and W. Wiseman, Jr. Continental shelf food chains of the Northern Gulf of Mexico. In: K. Sherman, L. Alexander and B. Gold (eds). Food Chains, Yields, Models

and Management of Large Marine Ecosystems. Boulder, Westview Press, 1991:67-106.

13. de Groot, R. Functions of Nature. Amsterdam, Wolters-Noordhoff, 1992.

14. Northridge, S. The environmental impacts of fisheries in the European community waters. A Report to the European Commission's Directorate General Environment, Nuclear Safety and Civil Protection. MRAG Ltd, 1991.

15. Goeden, G.B. Intensive fishing and a 'keystone' predator species: ingredients for community instability. Biol. Cons. 1982; 22: 273-281.

16. Skud, B.E. Dominance in fishes: the relation between environment and abundance. Science 1982; 216: 144-149.

17. Dayton, P.K. and M.J. Tegner. The importance of scale in community ecology: A kelp forest example with terrestrial analogs. In: P.W. Price, ed. A New Ecology: Novel Approaches to Interactive Systems, 1984.

18. Dayton, P.K. Competition, disturbance, and community organization: the provision and subsequent utilization of space in a rocky intertidal community. Ecol. Monogr. 1971; 41:351-389.

19. Simenstad, C.A., J.A. Estes, and K. Kenyon. Aleuts, otters and alternate stable-state communities. Science 1978; 200:403-411.

20. Paine, R.T. Food webs: linkage, interaction strength and community infrastructure. J. Animal Ecology 1980; 49: 667-685.

21. Castilla, J. and R. Duran. Human exploitation from the intertidal zone of central Chile: the effects on *Concholepas concholepas* (Gastropoda). Oikos 1985; 45:391-399.

22. Russ, G.R. and A.C. Alcala. Effects of intense fishing pressure on an assemblage of coral reef fishes. Mar. Ecol. Prog. Ser. 1989 56: 13-27.

23. Brock, V.E. and R.H. Rittenburgh. Fish schooling: a possible factor in reducing predation. J. Cons., Cons. Int. Explor. Mer 1960; 25: 307-317.

24. Fogarty, M.J., M.P. Sissenwine and E.B. Cohen. Recruitment variability and the dynamics of exploited marine populations. Trends in Ecology and Evolution 1991; 6(8):241-245.

25. Hofman, R. Cetacean entanglement in fishing gear. Mammal. Review 1990; 20:53-64.

26. Norse, E. Global Marine Biological Diversity. Washington, DC, Island Press, 1993.

27. Read, A. and D. Gaskin. Incidental catch of harbor porpoise by gill nets. J. Wildl. Mgmt. 1988; 52:517-523.

28. Agardy, T. Last voyage of the ancient mariner? BBC Wildlife Dec. 1992:30-37.

29. Choat, J.H. Fish feeding and the structure of benthic communities in temperate waters. Ann. Rev. Ecol. Sys. 1982; 13:423-429.

30. Reise, K. Long-term changes in the macrobenthic invertebrate fauna of the Wadden Sea: are polychaetes about to take over? Neth. J. Sea Res. 1982; 16:29-36.

31. Holme, N.A. Fluctuations in the benthos of the western English Channel. Oceanologica Acta SP, 1983: 121-124.

32. International Council for the Exploration of the Sea (ICES). Report of the study group on ecosystem effects of fishing activity. Unpublished report of the Study Group of ICES, 1992.

33. Johannes, R.E. and M. Riepen. Environmental, economic and social implications of the live reef fish trade in Asia and the western Pacific. Arlington, VA, The Nature Conservancy Report, 1995.

34. Sherman, K. Monitoring and assessment of large marine ecosystems: A global and regional perspective. In: McKenzie, ed. Ecological Indicators. Amsterdam, Elsevier Press, 1992.

35. Hatcher, B.G., R.E. Johannes, and A.I. Robertson. Review of research relevant to the conservation of shallow tropical marine ecosystems. Oceanogr. Mar. Biol. Ann. Rev. 1989; 27:337-414.

36. Sale, P. A reply. Trends in Ecology and Evolution 1990; 5: 25-27.

37. Peterson, M.N., ed. Diversity of Oceanic Life: An Evaluative Review. Center for Strategic and International Studies Significant Issues Series 1992; XIV(12).

38. Hilborn, R. and D. Ludwig. The limits of applied ecological research. Ecol. Appl. 1993; 3(4):550-552.

39. Mooney, H.A. and O.E. Sala. Science and sustainable use. Ecol. Appl. 1993; 3(4): 564-566.

40. Socolow, R. Achieving sustainable development that is mindful of human imperfection. Ecol. Appl. 1993; 3(4): 581-583.

41. Ludwig, D. Environmental sustainability: magic, science, and religion in natural resource management. Ecol. Appl. 1993; 3(4): 558-558.

42. Sinclair, M. Marine Populations. Seattle, WA, U. Washington Press, 1988.

43. Sherman, K. Can Large Marine Ecosystems be managed for optimum yield? In: K. Sherman and L. Alexander, eds. Variability and Management of Large Marine Ecosystems. AAAS Selected Symposium 1986; 99: 263-267.

44. Witman, J. and K. Sebens. Regional variation in fish predation intensity: a historical prespective in Gulf of Maine. Oecologia 1992; 90:305-315.

45. Ludwig, D., R. Hilborn, and C. Walters. Uncertainty, resource exploitation, and conservation: lessons from history. Science 1993; 260(2): 17; 36.

46. Rosenberg, A.A., M.J. Fogarty, M.P. Sissenwine, J.R. Beddington, and J.G. Shepherd. Achieving sustainable use of renewable resources. Science 1993; 262:828-829.

47. Pimm, S.L. The Balance of Nature. Chicago, U Chicago Press, 1991.

48. Johannes, R.E. Traditional conservation methods and protected marine areas in Oceania. In: J. McNeely and K. Miller, eds. National Parks, Conservation and Development. Smithsonian Institution Press, Washington, DC, 1984; 344-347.

49. de Fontaubert, C., D. Downes and T.S. Agardy. Protecting Marine and Coastal Biodiversity and Living Resources under the Convention on Biological Diversity. Washington, DC, Center for International Environmental Law, World Wildlife Fund and IUCN Publication, 1996.

50. Thorne-Miller, B. and J. Catena. The Living Ocean: Understanding and Protecting Marine Biodiversity. Wash., DC, Island Press, 1991.

51. Weber, P. Abandoned Seas: Reversing the Decline of the Oceans. Worldwatch Paper 166, Wash, DC, Worldwatch Institute, 1993.

52. World Resources Institute. The State of the World. Washington, DC, World Resources Institute, 1996.

53. GESAMP. The State of the Marine Environment. Oxford, Blackwell Scientific Publications, 1990.

54. Carlton, J. and A. Cohen. Alien Invasions in San Francisco Bay. USFWS Report 1996.

55. National Oceanic and Atmospheric Administration. Florida Keys National Marine Sanctuary: Draft Management Plan/ Environmental Impact Statement. Washington, DC, NOAA, 1995.

56. Agardy, T. Coral reefs and mangrove systems as bio-indicators of large scale phenomena: a perspective on climate change. In: Proc. of the Symp. on Biol. Indicators of Global Change 7-9 May 1992 Brussels, Belgium. J.J. Symoens, P. Devos, J. Rammeloo and C. Verstraeten, eds. Royal Academy of Overseas Sciences, 1994:93-105.

57. Markham, A. Potential impacts of climate change on ecosystems: a review of implications for policymakers and conservation biologists. Cli. Change Res. 1995 6:179-191.

58. Dayton, P.K. and M.J. Tegner. Bottoms beneath troubled waters: benthic impacts of the 1982-1984 El Nino in the temperate zone. In: P. Glynn, ed. Global Consequences of the 1982-1983 El Nino Southern Oscillation. Amsterdam, Elsevier Press, 1989:457-481.

59. Agardy, T. Prospective climate change impacts on cetaceans and implications for the conservation of whales and dolphins. Proceedings of the International Whaling Commission Scientific Committee Symposium on Climate Change and Cetaceans, 25-30 May 1996, Kahuku, Hawaii. 1996.

60. Levin, S. Science and sustainability. Ecol. Appl. 1993; 3(4).

61. Johannes, R.E. and J.W. MacFarlane. Traditional Fishing in the Torres Strait Islands. CSIRO, Hobart, Tasmania, 1991.

62. Johannes, R.E. Marine conservation in relation to traditional lifestyles of tropical artisanal fishermen. The Environmentalist 1984; 4(7).

63. National Research Council. Managing Troubled Waters: The Role of Marine Environmental Monitoring. National Academy Press. Wash., DC, 1990.

64. Fuentes, E.R. Scientific research and sustainable development. Ecol. Appl. 1993; 3(4):576-577.

65. Johannes, R.E., ed. Traditional Ecological Knowledge: A Collection of Essays. Gland, Switzerland, IUCN, 1989.

66. Knecht, R.W. 1990. Towards multiple use management: issues and options. In: S.D. Halsey and R.B. Abel, eds. Coastal Ocean Space Utilization. Amsterdam, Elsevier Science Publ., 1990.

67. Sorensen, J.C., S.T. McCreary, and M.J. Hershman. Coasts: Institutional Arrangements for Management of Coastal Resources. Renewable Resources Information Series, Coastal Management Pub. 1, 1984.

68. Elder, D. and J. Pernetta. Oceans: World Conservation Atlas. London, Mitchell Beazley Publishers, 1991.

69. Uravitch, J. A. Strategies for management and regulations of particularly sensitive sea areas: experience of the National Marine Sanctuaries Program of the U.S. Proceedings of the International Seminar of the Protection of Sensitive Sea Areas, Malmo, Sweden, 1990.

70. Eichbaum, W.M. and B.B. Bernstein. Current issues in environmental management: a case study of southern California's marine monitoring system. Coastal Management 1990; 18: 433-445.

71. Eichbaum, W.M., M.P. Crosby, M.T. Agardy, and S.A. Laskin. The role of marine and coastal protected areas in the conservation and sustainable use of biological diversity. Oceanography 1996; 9(1):60-70.

72. Hardin, G. The tragedy of the commons. Science 1966; 162: 1243-1248.

73. Berkes, F. Fishermen and the tragedy of the commons. Envir. Cons. 1985; 12(3):199-206

74. McNeely, J.A. Common property resource management or government ownership:Improving the conservation of biological resources. Intl. Rel. 1991; 10(3):211-225.

75. Doumenge, F. Human interactions in coastal and marine areas: present day conflicts in coastal resource use. Proc. on the Workshop of the Biosphere Reserve Concept to Coastal Areas, 14-20 August 1989, San Francisco, CA.

76. Agardy, T. Guidelines for Coastal Biosphere Reserves. In: S. Humphrey, ed. Proc. of the Workshop on the Application of the Biosphere Reserve Concept to Coastal Areas, 14-20 August, 1989, San Francisco, CA, USA. IUCN Marine Conservation and Development Report 1992, Gland, Switzerland.

77. Bliss-Guest, P. and A. Rodriguez. Sustainable development. Ambio 1987; 10(6):346.

78. Agardy, T. The Science of Conservation in the Coastal Zone: New Insights on How to Design, Implement and Monitor Marine Protected Areas. Proc. of the World Parks Congress 8-21 Feb. 1992, Caracas, Venezuela. IUCN, Gland, Switzerland. 1995.

79. Kimball, L. The Law of the Sea: Priorities and Responsibilities in Implementing the Convention. IUCN Marine Conservation and Development Report, Gland, Switzerland, 1995.

===== CHAPTER 3 =====

# GENERAL MEASURES TO ADDRESS THREATS TO MARINE ECOSYSTEMS

## SECTION 1. DEFINING LEVELS OF RESOURCE USE THAT ARE SUSTAINABLE

Effective conservation of the seas essentially entails controlling our use of natural resources so that the web of life needed to maintain the marine and coastal ecosystems which support assemblages of species (and humans as well) is not undermined. This is at the core of sustainability, the popular concept that embodies hope for the future. Sustainable use of resources, according to the Bruntland Commission where the phrase was coined and first used, is thought of as the use of resources today without compromising such use for future generations.[1] However, use of the term without stating precisely what is meant by it leads to confusion and potential conflict, as is discussed in the following pages.

Sustainable use of resources is now touted the world over as the solution to real and prospective global, regional and local environmental problems.[2] Yet, sustainable use as a concept is rarely defined; it is interpreted differently by ecologists, economists, sociologists, and politicians. The concept of sustainable use as it applies to marine resources is no better described than it is in terrestrial applications; in fact, it is made more complex by the difficulties inherent in working in an environment perceived as a global common, characterized by unclear jurisdictional boundaries and gaps in knowledge about its workings.

Sustainability, variously embodying the concepts of prolonged economic gain, ecologically sound development, low-level use of renewable resources and parity among all resource users in effecting management decisions, is sometimes seen as a panacea for a wide spectrum

*Marine Protected Areas and Ocean Conservation,* by Tundi Spring Agardy.
© 1997 R.G. Landes Company.

of environmental problems, local to global in scale. Politicians are quick to jump on the sustainability bandwagon, advertising their token interest in the environment to appease a worried public. Scientists and managers also contribute to the ubiquitous flag waving, hence the term "sustainable" is now a required component of conservation strategies and appears indiscriminately throughout the resource management literature.[3] But what exactly does sustainable use mean, and precisely how can it be attained?

Sustainable use is a concept that has rankled and divided ecologists, the circle in which it is discussed most often and from which it is borrowed and misused most aggreviously. The most commonly used meaning of ecological sustainability has to do with ecosystem function. For an activity to be sustainable within the functional limits of an ecosystem, that activity must not interfere with the workings of the system and its ability to keep critical parameters within homeostatic limits. That is, the activity must not cause environmental degradation in the systems sense. Removing organisms from an ecosystem or interfering with its critical processes can only be sustained over time if the system's functioning is not adversely impacted.

The simplest case of ecologically sustainable use is in the harvest of a non-living resource, where permissible levels of use are determined by the role of that resource in the ecosystem. Soil is such a resource, and its removal through use prevents organisms from using it as a basis for the development of complex and productive food webs. Soil removal is thus a non-sustainable resource use, unless the user gives something back to the system. In agriculture this added "something" is artificial fertilizer; however, even this replenishment does not necessarily lead to sustainability since energy inputs and outputs are not in balance.[4]

A slightly more complicated picture emerges in the harvest of a living resource. Here the level of take for a single species must not exceed a critical threshold, which is determined in part by the minimum number of individuals needed to replenish the population (often called the minimum viable population size.[5] This is the most common basis for quantifying levels of harvest which are sustainable, but it is a dangerously shallow interpretation of the concept of sustainability. Why? Because components of a living system do not exist in a vacuum without connection to other components. The population size of any organism in an ecosystem will influence the population sizes of other organisms, through direct predation, prey abundance, interference and competition. The organisms also influence the abiotic components of the environment in ways which are only now beginning to be explored. Thus, selective harvest of a population may influence the workings of the entire network that makes up the system; sustainable culling must be limited to levels which do not cause system decline and eventual collapse.

*Fig. 3.1. Commercial fishing boats, Scotland. Photo by T. Agardy.*

Even in the mechanistic realm of systems ecology this interpretation of the term sustainable use runs into problems, merely because the use of the word sustainable does not explicitly spell out a goal to be achieved. If the goal of resource managers is to maintain an ecosystem so that it functions exactly as it did before exploitation, then the level of use which is sustainable will likely be relatively low. We may harvest the "excess" when and only when it is available. If, however, the objective is to ensure that an ecosystem continues to operate, despite some changes to its structure and functioning, then permissible levels of use may be higher. In any case, since ecosystems do not exist as static, immutable entities but as dynamic and evolving systems, it will be difficult to approximate a level of use which has no effect on the ecosystem whatsoever. Despite this difficulty, this is the implicit, or, in some cases, explicit goal of ecologically sustainable use or development as it appears in many conservation strategies and resource management plans. This meaning creates an oxymoron of the term "sustainable development," since development or growth cannot occur without some effect on the environment.

The concept of a "limit of acceptable change" (LAC), which attempts to reconcile sociological and ecological viewpoints on limiting impacts on the environment, is sometimes haphazardly used in place of a sustainable limit. Although this term underscores a realistic assumption that human population growth and resource use will indeed affect the ecosystem, it attempts only to make more rigorous a problematic idea without necessarily addressing the problem. Acceptable to who? And for what time frame? These unanswered questions suggest the LAC idea is fraught with the same weaknesses as the sustainable use concept.

Economically viable strategies for harvesting living resources are not the same as ecological ones—in fact, they can directly counter ecologically-based strategies.[6,7] This applies in the macroeconomic sense: national assessments that are based on gross national products (GNP) and other standard valuations ignore the capital stock value of resources that are underutilized or not developed (whether for ecological reasons or not). The resources that do not find their way into the production stream are not valued in assessments of a nation's so-called economic worth, hence there is no reason whatsoever to protect them. From a microeconomic perspective, simple free market strategies that rely on the principles of supply/demand relations and discounted values dictate that use be kept to levels that optimize return on investment over some specific time horizon. As simplistic economic models are applied to renewable or non-renewable natural resource use, counterintuitive results may emerge.[8] The structure and functioning of the ecosystem is ignored in much of basic economic modelling, and where adequate substitutes for a product or use exist these may be targeted, as long as consumer demand is elastic. Substitution complicates the sustainable development picture and makes the planning of ecologically sustainable strategies less predictable.

There are other economic features which prevent the identified sustainable level of use from being realized. In some cases, developing a resource management plan which is economically enduring will require that levels be kept below what may be ecologically feasible; though the ecosystem might withstand greater levels of resource harvest, flooding the market acts to drive prices down and renders use economically unsustainable over the long-term. In other cases, just the opposite can be demonstrated; optimal and sustainable use over realistic time horizons will lead to resource depletion, since discounting changes the profit potential with time.

Another way in which economic and ecological sustainability may differ is in consideration of spillover effects. Where spillover is economically positive, increased use is encouraged. But from an ecological perspective, such spillover activities may interfere with the workings of the physical and biotic system, and may not be sustainable over time frames that are appropriate for ecosystem functioning. Again, analysis to determine what is sustainable and what is not occurs on different time scales in ecology and economics, so the two perspectives are rarely compatible.[9] An unnatural marriage may result from the forced melding of ecological and economic disciplines. And like many forced marriages, the relationship is rarely a happy or comfortable one.

While society places burdens on the scientist to provide evidence of "truth" in ecosystem functioning, it also has high expectations that economics will "prove" that human activity will have either a beneficial or an adverse impact in the long-term. Thus if the public swallows what the ecologist says about limiting use so it is sustainable, it is often up to the economist to show that it can indeed be sustained. There are other facets of this strained relationship. When economic arguments are used to justify environmental inaction, conservation is put at odds with economic growth.[10] Thus economists are put in the position of showing why limits to activity are economically beneficial on the one hand, and why unbounded use and ignorance of scientific evidence for management is warranted on the other.

For largely unexplained reasons, conservationists often treat sociological or ethical sustainability as being an independent but important component of sustainable use strategies.[11] In part this tendency is political because it lets people think that their individual well-beings are as much a consideration as economic viability and ecosystem health. This form of the sustainability is the sloppiest and most prone to misinterpretation; in the rare cases where the intended meaning is spelled out it is done so in vague and imprecise terms. In some discussions, sociological sustainability is by default taken to mean cultural preservation, which is an important objective but which has virtually nothing to do with the development of sustainable use strategies. Fair representation of all members of society in the management process is also sometimes called a sustainable strategy, and again the use of the term is strained. In the worst cases, the term is meant to describe an inalienable right to live one's life as one pleases: a meaning which has no place in discussions of resource management and is better left to discussions of cultural ethics. Its vagueness can be dangerous, since sustainability can then be distorted to mean that complete independence is justified, and presumably that every person is as free as the next to deplete resources to extinction.[12]

Can sustainable levels of use be determined for marine resources? The answer is yes; sustainability can be demonstrated in the use of marine resources, on small scales and in special cases.[13] Then why is marine resource use often not sustainable? Deceptively high standing stocks often lead to resource "mining," intensive and sustained mining then leads to exhaustion of the resource, extirpation of local populations of organisms, or even extinction of the species. Examples of this so-called rachet effect are extensively documented.[14] Another problem point is that yield calculations suggest a world in which ecosystem components exist independently of one another. Interspecific interactions are ignored–in fact, a static, immutable environment is an important assumption of the model.

The fact that many components of ecosystems are extensively connected means that renewable resource abundance is tied to a balanced system of exploitation (anthropogenic or otherwise). It also means that the ability of the resource to withstand culling, or the assimilative capacity of the system to withstand impacts, varies with overall ecosystem health. Thus when indirect environmental degradation undermines the physiological workings of a marine ecosystem, levels of harvest must be reduced to remain sustainable. So also with other stresses—sustainable use values must reflect changing environmental conditions, and planning must take into account the complete ecosystem.

Sustainable use of marine resources is often equated with fisheries projections of maximum sustainable yield (MSY). MSY is an ostensibly scientific parameter which is quantified through knowledge of the population dynamics of a species. It is usually generated from a population growth curve which is plotted against catch-per-unit-effort (CPUE). If the natural dynamic of a population is known over time, the optimal level of effort which maximizes return and does not impact population recovery is targeted. If fishing effort exceeds that magic maximum sustainable level, fishermen are faced with a diminishing return on investment. MSY models are thus used to buttress the argument that fishing is a self-regulatory activity.

What is wrong with MSY and why does its use not point us in a direction that ensures sustainable use?[15] One difficulty is that natural populations rarely exhibit the type of smooth growth that is a basic premise of the MSY calculation. In fact, the dynamic behavior of many fish and other commercially exploited species cannot be predicted with any certainty, so the presumed scientific basis for MSY calculations is often overexaggerated. Another problem point is that MSY calculations suggest a world in which ecosystem components exist independently of one another.[16] Interspecific interactions are ignored, in fact, a static, immutable environment is an important assumption of the model.

No resource exists independent of others in an ecosystem.[17,18] This extensive connectivity means that renewable resource abundance is tied to a balanced system of exploitation (anthropogenic or otherwise). It also means that the ability of the resource to withstand culling, or the assimilative capacity of the system to withstand impacts, varies with overall ecosystem health.[19,20] Thus when indirect environmental degradation undermines the eco-physiological workings of a marine ecosystem, levels of harvest must be reduced to remain sustainable. So also with other stresses: sustainable use values must reflect changing environmental conditions, and planning must take into account the complete ecosystem.[21]

In the MSY scenario, it is presumed that some "extra" amount of the resource exists for our taking. This yield can be increased if some other non-human consumer population decreases. For instance, severe overharvest of whales has dramatically reduced toothed whale populations in the North Atlantic, making one of their preferred prey species (capelin) more abundant. We can justify taking more capelin than we have historically because more of this extra abounds—we (or our pets, since capelin is most often processed to make cat and dog food) are lucky to eat what the unfortunate whales left behind.

Even this purportedly 'sustainable' use is risk-laden, however, since we are not as efficient consumers as the whales themselves. In fishing for capelin we waste by-catch, and in so-doing potentially alter the structure of the marine community. Although harvesting technologies may become more efficient, reducing estimated effort, the activity's harmful side effects may continue ecological degradation. Alterations in community structure and function thus may have undermined the stability of the marine ecosystem, such that further impact cannot be sustained over time. Again, what may be sustainable in theory may not be sustainable in practice because the theory is based on unrealistic assumptions.

The assumption that the preceding uses of the marine system are sustainable is predicated on the requirement of low level, non-invasive activity. When human populations grow and exert ever-greater pressures on resources, however, activities that were perceived as sustainable may lead to overexploitation and permanent ecosystem damage. These non-sustainable courses of action may be perceived as economically viable, but they compromise the ability of future generations to use the resources, directly countering the Brundtland Commission's definition of what is sustainable.[1]

In fact, high population growth coupled with diminished future value due to discounting rates could warrant harvesting resources to the point of extirpation, if a purely economic approach to sustainability is taken. Colin Clark (1988) has demonstrated how this can happen, using forestry as an example to conclude that clear-cutting can yield the greatest profit over virtually any time scale.[7] The complete decimation of forest resources is economically sustainable in the sense that the return on investment is maximized over any time period, but by any other interpretation the action is utterly unsustainable. Similar scenarios can be developed for living marine resources, including fisheries and even whaling. According to Clark, "The net present value criterion may justify extinction of a species," a result that is not only incompatible with ecological sustainability but its greatest nemesis.[22]

What about activities that reap natural resources but put something back as well? Aquaculture and mariculture, for instance, are ecologically and economically sustainable, so long as the fertilization of the system is done naturally and environmental stresses are minimized. However, problems arise from large scale mariculture activities as well. Among other things, such development threatens to diminish the functioning of natural coastal ecosystems the world over through habitat alteration. In Ecuador, for instance, large areas of mangrove forests have been destroyed to make shrimp ponds.[23] When mangrove deforestation occurred, nursery areas for the shrimp and other organisms may have been lost and recruitment of wild stock to seed ponds was restricted. This contributed to the near-collapse of the shrimp farming industry, dealing a hard blow to Ecuadorians and their investors. The lessons learned about ecosystem functioning and homeostasis, however, are invaluable.

Other mariculture-like activities can be sustainable as long as recruitment and other critical processes are left undisturbed. In the Sian Ka'an Biosphere Reserve in Mexico, for instance, spiny lobster fishermen are allocated tracts of the seabed for their own use by the local fishermen's cooperative.[24] Each fisherman builds small, portable artificial reef structures known as "cassitas" (little houses), which attract juvenile lobsters looking for shelter. Fishermen visit their plots several times a week and harvest adult lobsters that have grown up in the cassitas; they do not take immatures or gravid females. The fishery is sustained by a limited entry system that maintains a constant number of fishermen in the area. The sector is economically sustainable and, in fact, the per capita income of fishermen is the highest in the state by a wide margin.

The unique fishery in Sian Ka'an does not carry the guarantee of permanent sustainability, even if rules are maintained and entry is limited.

The reason for this is that other things affect lobster productivity, not the least of which is general environmental health. As water quality decreases in response to poor land use practices (including, for example, unmitigated hotel development that destroys mangrove forests and interrupts coastal processes, limestone mining operations that leach sediment into coastal waters and ship traffic that pollutes bays and offshore areas) ecosystem stress lowers productivity. Fishermen may then be surprised to find that their low-level and intrinsically sustainable patterns of resource use cannot be maintained. No living or non-living resource exists in a vacuum and all ecosystem components must be considered for use to be sustainable over many years or generations.

How then can effective coastal planning avoid potential disasters and increase the probability that activities can be sustained over the long-term? Through marine protected areas that conserve threatened resources and demonstrate key coastal management principles. These management principles can be shown to be effective in two ways: first, by establishing multiple use areas that allow for a variety of uses but which ensure that critical areas and processes are left undisturbed; and second, by making management flexible and responsive. In the Sian Ka'an example, essential knowledge about the ecological requirements of the lobster stocks can be used to develop more effective management plans. But these management plans must be adaptive, so that as environmental, economic and sociological conditions change, management can respond and also change.

Here we come to a central point: sustainable use, whether couched in economic, ecological, or socio-political terms, never implies a constant, immutable rate of harvest. Instead, the very essence of sustainability is its adaptability. This requirement for flexibility again means that critical processes must be understood, that ecosystem-wide responses must be taken into consideration and that the results of management be continuously monitored.[25,26] It also means that the best coastal management is that which is fine-tuned to local conditions, both in terms of ecosystem requirements and human needs. Such fine-tuning implies that management of human activity may be most effective when practiced on local scales. It is on these scales that socio-political sustainability (access to resources and their management) is also most likely.

The good news is that inherently low-level exploitation of living or non-living marine resources can be sustainable over the long-term. One such ecologically sound use is low-level recreational fishing, particularly when yield is kept intrinsically low due to limiting technologies, or extrinsically low through tag-and-release fishing. Other examples are provided by low-level artisanal fishing in which yield is again limited by technological limitations (e.g., free-diving sponge and conch fisheries, shellfish bull-raking). Bottom-up approaches to resource management are particularly successful when scientific information is made available to resource users (Mangel, 1985), and when local communities with vested interest in maintaining resource bases are given the responsibility and power to undertake environmental monitoring and evaluation.[27,28] Non-extractive uses of the marine ecosystem, including recreational boating and beach use, snorkeling and scuba diving, shipping, etc., are again all sustainable, but only as long as these activities do not interfere with the processes that maintain the ecosystem beyond critical threshold levels.

Most of us would admit clinging to a view that conservation is integral to wise economic development. However, by carelessly throwing about ill-defined goals of sustainability we dilute the strength and importance of this philosophy. Sustainability continues to carry different, often incompatible meanings.[11] However, the conflicting perspectives of what constitutes ecological, economic and socio-political sustainable use can be brought together. Doing so requires that objectives of sustainability are clearly spelled out. In coastal and marine areas, human activity is rarely sustainable because common property perceptions lead to competition for resources and a "take what you can" philosophy (Ray, 1988).[29] Notions of stewardship, where they exist, are being undermined by the exponential growth of coastal populations and the subsequent lowering of coastal environmental quality. If, however, management can demonstrate to users its ability to preserve resources for sustained use, then sociological and political interest will embrace sustainable planning.

If we do accept that ocean conservation is important, can we justify the expenses incurred in developing rational programs for its management? Here economics offers us little in the way of help, since conventional models of valuation are generally admitted to be deficient in estimating the net worth of the whole system.[30,31] The global ocean system is more than the sum of its quantifiable parts; and more than the estimated value of all fisheries stocks, shipping, oil and gas, recreational use, and other revenues combined. We have similar problems quantifying the value of tropical forests.[23] In addition to the value of the timber and other plant products they contain and the space they yield, tropical forests have important physiological functions, including preserving biodiversity, preventing erosion, landslides and local flooding, acting as a sink for carbon dioxide and preventing more rapid acceleration of the greenhouse effect, etc. Oceans are similarly vital in providing the negative feedback mechanisms needed for global homeostasis, including climate control, nutrient recycling and waste assimilation. These homeostatic functions are essentially unquantifiable, although recent economic models are becoming increasingly more sophisticated and able to deal with non-market values.

Coastal planning that is integrative and ecologically sustainable will pave the way for economic and socio-political sustainability, by virtue of the fact that these are all interconnected.[32] This management must

be fine-tuned to local needs and conditions and should rely on the principles of areawide multiple use planning. It must also be flexible to changing conditions, requiring a move away from traditional methods that define constant values for permissible exploitation, as well as continuous monitoring of management. In short, the pathway to sustainable use involves reconciling divergent views on what is sustainable and arriving at a clear statement of management goals, gaining better ecosystem knowledge and putting it to use, and development of management plans that are flexible, responsive and improved as better knowledge of ecosystems is gained.

The dynamics of the marine system are vast and complex.[33] Many marine species undertake long migrations or travel great distances during planktonic life stages.[34,35] This extends the physical range of processes and connections between habitats.[36] Marine systems are thus made especially complex by problems of scale, with nested temporal and spatial hierarchies invisible to the casual eye.

Planning management for sustainable use requires basic knowledge about the underpinnings of the system, as well as a thorough assessment of actual and potential uses of its components.[37] This is true especially because ecosystems are not static, unchanging entities—the final product of a long and ordered development—but rather a complex and dynamic web of interactions which fall prey to cumulative impacts.[38] Complete knowledge of the dynamics and the full effect of impacts is neither a necessity nor a realistic possibility. However, management that is based on rudimentary knowledge can be fine-tuned as more knowledge is gained.[39]

Many policy makers and administrators view management of ocean use for sustainability an insurmountable task, due to scaling and dynamics as well as complicated jurisdictions. But we rarely balk at managing terrestrial use of resources, despite incomplete knowledge and pervasive conflicts. Despite important ecological differences between marine and terrestrial systems, the objectives in achieving sustainable use are the same. Those objectives, and the time scales in which they are to be reached, must be spelled out in order to accommodate all points of view.

Oceanic and coastal systems are ripe targets for innovative management, since we are not yet prejudiced by established convention on how to regulate use.[40] Tax incentives, pollution surcharges and other measures can be used to provide economic incentives for limiting use to sustainable levels. Systems should be dealt with in a systematic way, with ecology providing the underpinning for economic rationalizations.

Multiple use planning which takes into account the various sensitivities of living resources, the processes which replenish them, and the interactions between abiotic and biotic components, and that accounts for historic, real-time and prospective impacts can lead to sustained and ecologically sound use. For marine resources, much room exists for growth and development, and growth potentials can be maximized without comprising the ocean environment. Resource economics, acting in conjunction with ecosystem ecology, may be the best equipped of all the disciplines to suggest long-term strategies to protect the environment.[29] This may seem like a non-conservative and strange attitude for avowed conservationists, but development and conservation are not mutually exclusive activities.[2]

In an ideal world, a "hands off" blanket policy for the oceans might be favored. But we are faced with the real world, one that is near bursting at the seams–where many of the seams happen to be at the land/sea interface. Poor management can only lead to failure, given anticipated future pressures and a long history of chronic negative impacts on critical processes. On the other hand, with management that is holistic in assessing systems and prospective uses, there is no reason to believe that conservation objectives cannot be met. There is a sustainable world out there for the making, if only we could agree on what we mean by it.

## SECTION 2. IDENTIFYING AND PROTECTING AREAS CRITICAL TO ECOSYSTEM FUNCTION

We are beginning to see signs of a marine revolution of sorts, with changing attitudes, more serious attention to conflicts and prospective problems and the employment of new models and methods for managing our impacts on coastal and open ocean ecosystems. Closed areas, harvest refugia and multiple use marine protected areas are increasingly being selected from the portfolio of management options available to resource managers, largely because conventional methods have repeatedly failed. This failure is now beginning to enter the realm of public consciousness as mismanagement has begun to affect fishermen, other users of ocean resources and space and consumers alike. Closed areas and multiple use marine protected areas are two tools that move marine management away from largely ineffective sectoral control to true conservation that benefits humans and nature.

The recent popularity of these methods has in part to do with new scientific developments that reflect where we currently stand in our knowledge of populations dynamics, behavior of ecological forcing functions and the identity and location of critical processes.[41] The link between well-protected coastal areas and the maintenance of marine fisheries resources has been well-documented.[42] Ecologically critical processes in nearshore ecosystems are often concentrated in areas that can be easily demarcated by physical parameters such as reef formations, extensive banks or other shallow areas, certain types of coastal wetlands, continental shelf breaks and frontal systems, etc.[43,44] These areas and the critical processes they support, such as fish spawning, migratory pathways, breeding, settlement and concentrated feeding, inter alia, can be effectively protected at relatively little opportunity or direct cost, through small but well-enforced harvest refugia or marine protected areas.

In the new generation of marine conservation, critical areas can be protected as small but discrete "core areas." In ultradynamic systems such as high latitude shelf systems, cores may have to be seasonally or spatially variable to match the ecological dynamics. If the science is adequate, well-protected core areas need not be large and need not preempt continued human use of ocean space. Instead, a suite of small but strictly protected core areas will ensure that the ecosystem's self-sustaining processes are protected, guaranteeing further production and providing the basis for continued sustainable use.[45,46] For marine populations with recruitment from afar, protection measures will have to reflect ecological reality with cores in areas far from the original habitat requiring conservation attention.[47,48]

## SECTION 3. USING ZONING TO RESOLVE USER CONFLICTS

As human populations in coastal areas increase and access to marine resources and space grows through technological advance and greater necessity, the potential for conflicts between users of marine resources increases. Users compete among themselves for the same stock of resources, resources that are present in ever-diminishing supply. These intra-sectoral conflicts can and often do lead to a rush to overexploit in the face of competition for scarce resources. However, many uses are intrinsically incompatible, and even low intensity use can lead to severe or irreconcilable conflicts. For instance, fisheries utilizing dragged gear (purse seines, trawls, etc.) conflict with fisheries using fixed gear (weirs, fish pots, etc.). Coastal tourism development is often at odds with fisheries development or mining. Industrial or port development can bring local communities into conflict with development agencies.

Even nature-based tourism or ecotourism is not a panacea, and conflicts can emerge when the differing needs and expectations of all users of the coastal area are not anticipated. Some of the problems that can lead to incompatibility of uses include: 1) tourism development attracts attention to the resource, thereby creating ever-greater demand for access and use; 2) tourism can create an elitist situation, providing access only to those who can afford it; 3) inappropriate tourism development can disrupt self-regulating traditional systems of use; 4) continuing tourism development can create incentives for local population expansion, leading to increased resource use pressures; 5) tourism, when coupled to the formation of a recreational use-oriented marine park, can lead to an increase in overexploitation outside boundaries of core areas; 6) successful ecotourism ventures can create a false sense of security that coastal management issues throughout the country or region are being dealt with effectively; 7) tourist attractions can deflect attention away from areas where conservation is most needed, since such development and the parks that often support them are most easily established in conflict-free, relatively pristine areas; 8) economic sustainability will take precedence over ecological sustainability; 9) the presence of foreign tourists can create a situation where foreign value systems override local value systems; and 10) marine tourism activity can open the door for alien species introductions that subsequently undermine natural biodiversity.

In what may be a new era of marine conservation, new perspectives in landscape and systems ecology provide the groundwork for zoning plans that will protect vital areas and move us closer to sustainability. Multiple use areas that are integrative, scientifically based and realistically planned will allow continued exploitation of marine and coastal resources and accommodation of potential users without conflict.

Multiple use zoning is a technique that marine parks planners are being to use to resolve real or prospective use conflicts, allowing the protection of sensitive or critical areas and allowing utilization of marine resources that is sustainable over the long term. Such multiple use protected areas are based on ecology: core areas define such critical areas as feeding grounds, spawning areas, or migration corridors. Buffer zones of regulated use surround these sensitive areas, ensuring that human use of the ecosystem will not degrade critical areas, neither directly nor indirectly. The outer boundaries of the multiple use marine protected areas signify the limits inside that interagency and user cooperation is maximized. Since multiple use areas are planned with the consultation and participation of local users, the users themselves become stewards of the area, become responsible for day-to-day management and receive the benefits of a thriving, productive ecosystem.

As stated repeatedly throughout this book, the nature of a multiple use protected area, its design and its regulatory framework will all depend on the primary objectives it helps to achieve. In some places, conservation will be the prime motivating force for a protected area designation; in others, preservation of traditional use will be paramount and so forth. The objective(s) identified will influence the size, shape and other design constraints of the protected area, and its implementation. Since specific objectives and circumstances vary so widely in coastal areas around the world, no model for coastal protected areas can be said to be universally applicable.[49] However, guidelines that frame the most important considerations do exist to facilitate the decision-making and implementation process.

The Great Barrier Reef Marine Park (GBRMP) is without question the largest, most ambitiously planned and most highly praised multiple use marine protected area in the world. Its vast expanse reaches from far north Queensland, where the Torres Strait separates the Australian continent from Papua New Guinea, to the southern margin of the reef, some 1600 kilometers to the south. In total, the GBRMP covers the largest single collection of coral reefs and associated habitats in the world. The Cairns Section (Fig. 3.1), one of four sections of the GBRMP covering an area of 35,000 km$^2$, is one of the Reef areas most heavily used by tourists and one of the most well-studied areas on the reef tract.[44]

The GBRMP was established in 1976 through an Act of the Australian Parliament. An independent Commonwealth statutory body called the Great Barrier Reef Marine Park Authority (GBRMPA) oversees permitting enforcement, research and zoning amendments (currently called for by law every five years). Senior administrators of the Authority are based in the national capital, Canberra, while day-to-day management is accomplished by headquarters staff in Townsville and outposted staff at various mainland and island locales. The Cairns Section was proclaimed in November 1983, after careful review of zoning plans for the northern Reef area.

The primary objective that the forward-looking GBRMP aimed to achieve through its multiple use zoning plans was to accommodate anticipated growth in coastal and marine tourism while maintaining environmental quality and avoiding conflict with other economic sectors. The Authority believes, "any use of the Reef or associated areas should not threaten its existing essential ecological characteristics and processes," and adopts as the primary goal of GBRMP management "to provide for the protection, wise use, understanding and enjoyment of the Great Barrier Reef in perpetuity."[50] Thus the park, and in particular the Cairns Section, was zoned first and foremost to accommodate tourism. This is in marked contrast to some other marine protected areas, since traditional use of Great Barrier Reef space and resources was only a secondary, and largely minor, consideration.

Marine-based development of the Great Barrier Reef area accounts for revenues exceeding $1 billion annually. Yet despite its obvious successes and high-profile image, the GBRMP may be somewhat misleading as a model for multiple use marine protected area planning. This is the case first and foremost because no parallel exists anywhere else in the world; no other biologically important coral reef system is so vast, so pristine, so far from centers of anthropogenic impact and so amenable to multiple use management in being big enough to accommodate most users without conflict. Thus, there may be constraints limiting the usefulness of the GBRMP model for application in other areas. And while the Australian government has been wise in capitalizing on tourist interest, future growth even in the highly managed GBRMP may be cause for alarm. Many of the public comments that emerged from the re-zoning hearings following the implementation of the Cairns Section's initial zoning plan expressed concern that development, and particularly tourism development, was being promoted by GBRMPA at the expense of conservation. Friction among shareholders and differing

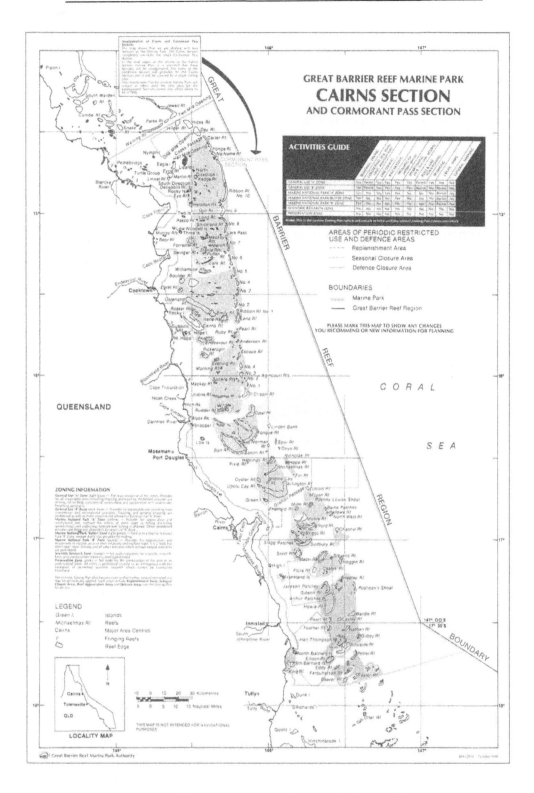

*Fig. 3.2. The Cairns Section of the Great Barrier Reef Marine Park, Australia.*

expectations among users of the Reef may undermine the Authority's ability to manage its marine area wisely. Lastly, the inability of the GBRMPA to control or even influence land use in Queensland has meant that land-based sources of pollution entering the Great Barrier Reef Lagoon have greatly degraded large parts of this ecosystem.[51]

Thus, well-planned marine protected areas seek to preserve ecosystem integrity while accommodating user groups to the maximum extent possible. Our success in implementing such protected areas is hindered, however, by gaps in understanding about marine ecology in general and in what levels of use of resources are in fact sustainable. This paucity of knowledge underscores the great need for marine protected areas in overall regimes to manage and conserve the coasts and seas. Beyond providing a sense of place and a geographical framework for local management, marine protected areas can contribute in two major ways to protection of marine systems. They allow decision-makers to buffer against unforeseen management mistakes (which are sadly all too frequent) and they provide a framework for testing management measures so that conservation and management can be undertaken in efficacious ways.

## SECTION 4. ESTABLISHING OWNERSHIP AND LAYING THE GROUNDWORK FOR STEWARDSHIP

The development of marine protected areas for the purpose of conserving marine biological diversity is a worthwhile endeavor for any nation wishing to derive maximum benefit from its natural heritage. However, while planning can be orchestrated by national government, the involvement of users of the resources will make the process more effective.[52,53] The user groups should not be thought of as the recipients of a plan or the objects of regulation, but rather as partners with government who can share the burden of responsibility for both planning and implementation of conservation measures.[54] For this reason, coastal communities and other users of marine resources and ocean space should be brought into the planning process right from the start.[49,53] This starting point will often be in setting specific goals that portions of the strategic plan will target.

The involvement of user groups, who are essentially stakeholders in the conservation process, is crucial for three discrete reasons. First, stakeholders have traditional knowledge about resources and ecosystems that will be important to determining levels of sustainable use, prioritizing conservation needs and selecting conservation measures and developing management regimes that are adaptive. Second, stakeholder involvement can strengthen the existing basis for the conservation ethic in societies and create the necessary conditions for fostering stewardship.[55] Stewardship for the environment encompasses three elements: use (and with it recognition of value of resources), responsibility and protection. Without such fundamental stewardship, even the best laid plans for conservation of biological diversity will fall on deaf ears and ultimately fail. The third reason that stakeholder participation is critical to the effective conservation of biological diversity is that those who have "bought in" to the planning process will be more amenable to sharing responsibility for both management of resources and monitoring conservation progress.[54,56]

Involvement of local people will be most beneficial to the planning process when balance can be achieved between analysis that occurs at the national scale and involvement that occurs at the local scale. This balance will be achievable when those with responsibility for resource management and environmental protection work closely with those who are affected by regulation, those who control the instruments for environmental problem-solving and those with relevant information and expertise.[57]

## SECTION 5. SPECIAL MANAGEMENT FOR SPECIAL SPECIES

Special protection is often needed for endangered species threatened with extinction, or for highly vulnerable habitat types. General management measures must take these needs into account, and marine protected areas provide an important (and in some cases the only) means to protect critical habitats for endangered species or especially sensitive habitats/species assemblages.[45,58-60] Indeed, if marine protected areas are a concept or tool that has penetrated the minds of the general public it is usually in this context.

Species of special concern are those that are highly threatened and face imminent danger of extinction, as well as those that play a central role in ecosystem functioning. Conservation groups have had a long history of working to protect threatened species or aid in the recovery of endangered populations, and most of this work has had a solid grounding in the sciences of conservation biology and population dynamics.

Species that most merit conservation attention are generally those that are most highly threatened or endangered. However, species of special concern may include those organisms that play a central role in ecological functioning and/or conservation itself. Species that may serve as focal points for conservation efforts and may provide an important impetus for the establishment of marine protected areas include: 1) species in imminent danger of extinction, especially those to which insufficient attention is being given to their recovery; 2) species that play a central role in ecological communities and may serve as important indicators of the condition of habitats or that may signal declines in an ecosystem, including organisms known as "keystone species" and "umbrella species;" and 3) species that serve as a conservation "hook," allowing us to scale up from single species-directed conservation efforts to broader conservation of entire systems, including organisms that readily capture public attention and can be utilized for raising awareness.

Strategies for conserving threatened or endangered species must be rooted in solid conservation biology that takes into account the ecological requirements, population replacement rates and factors affecting population recovery for the target species. The development of action plans, management plans and recovery plans for certain species of special concern, and protected area regimes in which to implement such plans are needed.

Sea turtles are particularly important in galvanizing attention to marine conservation problems and have provided the focus for the establishment of many marine protected areas in tropical coastal regions. Figure 3.2 shows a green turtle, *Chelonia mydas*, in its underwater habitat. All eight species of sea turtle found worldwide are threatened or endangered. The conservation of sea turtles provides great opportunities to catalyze action by leveraging the attention given these organisms, to serve as a starting point or "hook" for wider ecosystem-scale coastal and marine conservation projects and to act as umbrella species for coastal systems inclusive of land-sea linkages.

It must be said, however, that the days of using protected areas solely to safeguard single species or highly threatened species assemblages are past. These days marine protected areas may use species protection as a starting point for establishing more comprehensive and ecologically realistic systems of management.[61] And species protection does figure very prominently in the development of zoning plans, where critical habitats can be strictly protected in specially demarcated zones within a protected area.

## SECTION 6. INTEGRATING MANAGEMENT ACROSS ALL SECTORS

To be truly effective, coastal management should address, or at least consider, four general conservation goals in a fully integrated fashion.[62] First, where pressures to exploit resources are high, coastal management should adhere to a regime that is based on scientifically sound, rational definitions of what levels of use are truly sustainable. These sustainable use levels must be identified with the entire system in mind, not a single species at a time. Second, integrated coastal management should confer special management attention to those components of the ecosystem (species or processes) that are highly threatened. Endangered species such as endemic fishes, marine mammals, sea turtles, coastal forests, etc. are useful indicators of general trends in environment condition and require special management. Third, coastal management should, wherever possible, utilize some type of zoning system to protect habitats that act as critical areas for the ecosystem in question. Even low diversity areas with functional significance to the ecosystem should receive priority protection. Lastly, specific conservation

*Fig. 3.3. Green turtle: a marine endangered species. Photo by J. Spring.*

measures, since they do not occur in a vacuum but rather exist in the context of a wider matrix of differing management regimes, must tackle the problems of indirect degradation of the target ecosystem. They should act to focus scientific and political attention on the indirect degradation of target areas through point source pollution, uncontrolled run-off and poor watershed management, inconsistent coastal use outside the managed area and global change. Only in this way can integrated coastal management be truly comprehensive and integrative.

Coastal zone management in the U.S. and in some other industrialized countries has emphasized the restructuring of administrative frameworks to make control of all marine/coastal resource use and space as coordinated as possible.[63] While such administrative coordination is an important and laudable goal, coastal management in many other parts of the world requires more than amending extant regulatory systems. Effective conservation of coastal ecosystems worldwide requires the following: 1) deciphering and articulating people's needs and thus their potential reliance and impact on coastal systems; 2) using the best available scientific information to determine what levels and what kinds of resource use are optimal; and 3) developing a management structure that accommodates a wide range of needs with minimum user conflict and empowers local communities to be wise stewards of the seas.[64]

Elaborating specific objectives for coastal management, whether under the aegis of a national/provincial/state coastal management plan or a multiple use protected area, forces objective evaluation of people's requirements, expectations and future impacts on coastal systems.[65] The importance of canvassing all users in all segments of society to get at this goal-setting cannot be overstated. Stakeholders can be found in various levels of government, businesses and local communities who may depend on coastal resources for subsistence or market economies.[53] Other, less directly linked stakeholders include non-governmental organizations and multilateral aid agencies or other investors. All of these stakeholders have interests–sometimes conflicting–that must be considered in the development of a realistic and equitable coastal management plan.

We must also harness the science we currently have available to us to develop rigorous management, to in turn ensure that use of marine resources and ocean space will be sustainable over long time frames. This science should draw from advances in the fields of ecology, population dynamics/genetics, physical oceanography and hydrology and environmental studies.

However, it would be foolish to think we can develop a generic model for integrated coastal management, even one based on the best available science, that will take root wherever we plant it. Political, cultural and economic differences around the world will mean not only that objectives will vary from place to place but also that the mechanisms by which those goals are reached will likewise be diverse. Adapting management so that it is effective requires a thorough understanding of local conditions, sociological as well as environmental. Top-down approaches aimed at developing large scale frameworks for coastal management must be complemented by bottom up approaches that root this management in local realities.

Integrating management requires the establishment of systems of governance that facilitate decision-making across all linked sectors, in order to promote use of natural resources that is ecologically and economically sustainable. Decision-making should be guided by a set of generally accepted principles (e.g., Rio Declaration of Principles, FAO Fisheries Code of Conduct), be integrated in nature and have the capacity to craft sustainable development. The integration necessitates identification of existing and projected uses and their interactions, and in so doing promotes compatibility and balance of uses. Integrated management of coastal and marine resources and space should not aim to replace sectoral management but supplement it, ensuring that coordination leads to efficiency and conflict resolution or avoidance. Geographically speaking, the integration of management and coordination of governance should extend to the five zones that are affected by and in turn affect marine biodiversity: inland areas, coastal plains, coastal or nearshore waters, offshore continental shelf waters (usually within EEZ jurisdictions) and high seas areas. Practically speaking, however, integration of national policies for management will focus on watersheds, coastal lands and waters and offshore waters within national jurisdiction.

REFERENCES

1. Brundtland, G. Our Common Future. In: V. Martin, ed. For the Conservation of the Earth. Golden, Colo. Fulcrum Pres, 1988:8-12.
2. IUCN, UNEP and WRI. Caring for the Earth. Gland, Switzerland, IUCN, 1991.
3. Levin, S. Science and sustainability. Ecol. Appl. 1993; 3(4).
4. Goldemberg, J., T.B. Johansson, A.K.N. Reddy, and R.H. Williams. Energy for a Sustainable World. Washington, DC, World Resources Institute, 1989.
5. Gilpin, M. and M. Soule. Minimum viable populations:processes of species extinction. In: M. Soule, ed. Conservation Biology:The Science of Scarcity and Diversity. Ann Arbor, U. Mich. Press, 1986:19-34.
6. Tisdell, C. Sustainable development:differing perspectives of ecologists and economists, and relevance to LDCs. World Development 1988; 16(3):373-384.
7. Todd, J. An ecological economic order. Fifth Annual E.F. Schumacher Lecture, Cambridge, MA:Harvard University, 1985.
8. Clark, C. Clear-cut economies. The Sciences, NY Academy of Science, Winter 1988.

9. Holden, C. Multidisciplinary look at a finite world. Science 1990; 249:18-19.
10. Carnegie Commission On Science, Technology, and Government. E Cubed:Organizing for Environment, Energy, and the Economy in the Executive Branch of the U.S. Government. Washington, DC. Task Force on Environment and Energy, 1990.
11. Salwasser, H. Sustainability needs more than better science. Ecol. Appl. 1993; 3(4):587-589.
12. Agardy, T. Accommodating ecotourism in multiple use marine reserves. Ocean and Coastal Management 1993; 20:219-239.
13. Bliss-Guest, P. and A. Rodriguez. Sustainable development. Ambio 1987; 10(6):346.
14. Caddy, J.F. and G.D. Sharp. An Ecological Framework for Marine Fishery Investigations. FAO Fish. Tech. Pap. 1983; 283
15. Larkin, P.A. An epitaph for the concept of MSY. Trans. Am. Fish. Soc. 1977; 107:1-11.
16. Caddy, J.F. Species interactions and stock assessment–some ideas and approaches. In: C.Bas, R. Margalev and S.P. Rubies, eds. Simposio Internacional sobre des areas de afloramiento mas importantes del Oeste Africano (Cabo Blanco y Benguela). Instituto de Investigaciones Pesqueras, Barcelona, 1985.
17. Paine, R.T. Food web analysis through field measurement of per capita interaction strength. Nature 1992; 355:73-75.
18. Jones, P.J. A review and analysis of the objectives of marine nature reserves. Ocean and Coastal Management 1994; 24:149-178.
19. Barrett, C.W., G.M. Van Dyne and E.P. Odum. Stress ecology. Bioscience 1976; 26(3):192-194.
20. Holling, C.S. Adaptive Environmental Assessment and Management. New York, John Wiley and Sons, 1988.
21. Dayton, P., R. Hofman, S. Thrush, and T. Agardy. Environmental effects of fishing. Aquatic Conservation: Marine and Freshwater Ecosystems 1995; 5:205-232.
22. Norse, E. Global Marine Biological Diversity. Washington, DC, Island Press, 1993.
23. de Fontaubert, C., D. Downes and T.S. Agardy. Protecting Marine and Coastal Biodiversity and Living Resources under the Convention on Biological Diversity. Washington, DC, Center for International Environmental Law, World Wildlife Fund and IUCN Publication, 1996.
24. Miller, D. Learning from the Mexican Experience:Area apportionment as a potential strategy for limiting access and promoting conservation of the Florida lobster fishery. In: K. Gimbel, ed. Limiting Access to Marine Fisheries:Keeping the Focus on Conservation. Washington, DC, Center for Marine Conservation and World Wildlife Fund, 1994.
25. Agee, J.K. and D.R. Johnson. Ecosystem Management for Parks and Wilderness. Seattle, WA, U.Washington Press, 1988

26. Mangel, M., L.M. Talbot, G.K. Meffe, M.T. Agardy, D.L. Alverson, J. Barlow, D.B. Botkin, G. Budowski, T. Clark, J. Cooke, R.H. Crozier, P.K. Dayton, D.L. Elder, C.W. Fowler, S. Funtowicz, J. Giske, R.J. Hofman, S.J. Holt, S.R. Kellert, L.A. Kimball, D. Ludwig, K. Magnusson, B.S. Malayang, C. Mann, E.A. Norse, S.P. Northridge, W.F. Perrin, C.Perrings, R.M. Peterman, G.B. Rabb, H.A. Regier, J.E. Reynolds III, K. Sherman, M.P. Sissenwine, T.D. Smith, A. Starfield, R.J. Taylor, M.F. Tillman, C. Toft, J.R. Twiss, Jr., J. Wilen, and T.P. Young. Principles for the conservation of wild living resources. Ecol. Appl. 1996; 6(2):338-362.

27. Mangel, M. Decision and Control in Uncertain Resource Systems. New York, Academic Press, 1985.

28. Rubenstein, D.I. Science and the pursuit of a sustainable world. Ecol. Appl. 1993; 3(4):585-587.

29. Ray, G.C. Sustainable use of the ocean. In: Changing the Global Environment. New York, Academic Press, 1988:71-87.

30. Hardin, G. Paramount positions in ecological economics. In: Ecological Economics: The Science of Sustainability. R. Costanza, ed. Environmental Economics, 1991:47-57.

31. Hoagland, P., Y. Kaoru and J.M. Broadus. A methodological review of net benefit evaluation for marine reserves. World Bank Environment Dept., Pollution and Environmental Economics Division. Env. Econ. Ser. 1996; 26.

32. Burbridge, P.R., N. Dankers and J.R. Clark. Multiple use assessment for coastal management. Coastal Zone 1989; 89:33-45.

33. Bernal, P. and P.M. Holligan. Marine and coastal systems. Proceedings of the International Conference on an Agenda of Science for Environment and Development into the 21st Century, 24-29 Nov 1991 Vienna: Section II, Theme 8:1-10.

34. Ayers, J. Population dynamics of the marine clam, *Myaarenaria*. Limnology and Oceanography 1956; 1:26-34.

35. Bohnsack, J.A. and J.S. Ault. Management strategies to conserve marine biodiversity. Oceanography 1996; 9(1):73-82

36. Bakun, A. Definition of environmental variability affecting biological processes in large marine ecosystems. In: K. Sherman and L. Herander, eds. Variability and Management of Large Marine Ecosystems, Washington, DC AAAS Press, 1986:89-107.

37. Costanza, R., W.M. Kemp and W.R. Boynton. Predictability, scale, and biodiversity in coastal and estuarine ecosystems:implications for management. Ambio 1993; 22(2-3):88-96.

38. Gosselink, J.G. and L.C. Lee. Cumulative impact assessment in bottomland hardwood forests. Center for Wetland Resources, Louisiana State University, 1987. LSU-CEI-86-09.

39. Halbert, C.L. How adaptive is adaptive management? Implementing adaptive management in Washington State and British Columbia. Reviews in Fisheries Science 1993; 1(3),261-283.

40. Agardy, T. Advances in marine conservation:the role of protected areas. Trends in Ecology and Evolution 1994; 9(7):2676-270.

41. Hatcher, B.G., R.E. Johannes, and A.I. Robertson. Review of research relevant to the conservation of shallow tropical marine ecosystems. Oceanogr. Mar. Biol. Ann. Rev. 1989; 27:337-414.

42. Caddy, J.F. Toward a comparative evaluation of human impacts on fishery ecosystems of enclosed and semi-enclosed seas. Rev. Fish. Sci. 1993; 1(1):57-95.

43. Dearden, P. Protected areas and the boundary model:Mears Island and Pacific Rim National Park. Geographica 1988; 32(3):256-265.

44. Woodley, S. 1989. Management of water quality in the Great Barrier Reef Marine Park. Water Science and Technology 21(2):31-38.

45. Kelleher, G. Identification of the Great Barrier Reef region as a particularly sensitive area. In: Proc. of the International Seminar of the Protection of Sensitive Sea Areas, 1990:170-179.

46. Vermeij, G.J. When biotas meet: understanding biotic interchange. Science 1991; 253:1099-1104.

47. Lopez, J.M., A.W. Stoner, J.R. Garcia and I. Garcia-Muniz. Marine food webs associated with Caribbean Island mangrove wetlands. Acta Cientifica, 1988.

48. Pineda, J. Predictable upwelling and the shoreward transport of planktonic larvae by internal tidal bores. Science 1991; 253:548-550.

49. Agardy, T. Guidelines for Coastal Biosphere Reserves. In: S. Humphrey, ed. Proc. of the Workshop on the Application of the Biosphere Reserve Concept to Coastal Areas, 14-20 August, 1989, San Francisco, CA, USA, IUCN Marine Conservation and Development Report 1992, Gland, Switzerland.

50. Great Barrier Reef Marine Park Authority. Rezoning plan for the Cairns Section. GBRMPA, Townsville, 1992.

51. Bell, P.R.F. and I. Elmetri. Ecological indicators of large scale eutrophication in the Great Barrier Reef lagoon. Ambio 1995; 24(4):208-215.

52. Asava, W. Local fishing communities and marine protected areas in Kenya. Parks 1994:4(1):26-34.

53. Lees, A. Traditional ownership, development needs and protected areas in the Pacific. Parks 1994; 4(1):41-47.

54. Fiske, S.J. Sociocultural aspects of establishing marine protected areas. Ocean and Coastal Management 1992; 18:25-46.

55. White, A.T., L.Z. Hale, I. Reynard and L. Cortesi. Collaborative and Community Based Management of Coral Reefs. West Hartford, CT, Kumarian Press, 1995.

56. Gubbay, S. Marine Protected Areas: Principles and Techniques for Management. London, Chapman and Hall, 1995.

57. World Bank. World Bank Participation Sourcebook. Environment Department, Social Policy and Resettlement Policy Division, Participation Series Working Paper 1996; 19.

58. Massin, J.M. Waste disposal at sea in light of sensitive marine areas concepts. In: Proceedings of the International Seminar on the Protection of Sensitive Sea Areas 1990:337-349.

59. Peet, G. Protection of vulnerable areas in different geographic and ecological situations. In: Proceedings of the International Seminar on the Protection of Sensitive Sea Areas, 1990:240-253.

60. Voskresensky, K. 1990. Intergovernmental work in the maritime vulnerable areas. In: Proceedings of the International Seminar on the Protection of Sensitive Sea Areas, 1991:6-16.

61. McClanahan, T. Are conservationists fish bigots? Bioscience 1990; 40(1):2.

62. Humphrey, S.R. and B.R. Smith. A balanced approach to conservation. Cons. Biol. 1990; 4(4):341-343.

63. Eichbaum, W.M. and B.B. Bernstein. Current issues in environmental management:a case study of southern California's marine monitoring system. Coastal Management 1990; 18:433-445.

64. Eichbaum, W.M., M.P. Crosby, M.T. Agardy, and S.A. Laskin. The role of marine and coastal protected areas in the conservation and sustainable use of biological diversity. Oceanography 1996; 9(1):60-70.

65. Knecht, R.W. 1990. Towards multiple use management:issues and options. In: S.D. Halsey and R.B. Abel, eds. Coastal Ocean Space Utilization. Amsterdam, Elsevier Science Publ., 1990.

# PART II:
# COUNTERING THREATS TO OCEAN SYSTEMS THROUGH MARINE PROTECTED AREAS

# CHAPTER 4

# An Introduction to Marine Protected Areas and Management Goals

## SECTION 1. MARINE PROTECTED AREAS AS A TOOL FOR MANAGEMENT

Conservationists have only focused their attention on marine issues in recent times, therefore protection and management of marine resources lag several decades behind the land-based environmental movement. There are many obstacles impeding successful conservation of marine systems, many of which were described in earlier chapters of this book. The fluid nature of the environment and the nebulous character of ecological boundaries account for some of these difficulties and have made it necessary for conservation biologists to develop new models (Fig. 4.1). Truly effective marine conservation requires that we give up our traditional preoccupation with conserving structure (by erecting fencing around the fragments of systems that we feel have a structure worth protecting) and instead direct ourselves towards safeguarding the critical ecological processes that are responsible for maintaining that valuable structure. Though such a functional approach is not *unique* to marine conservation it is the only feasible option for relieving some of the pressures that a burgeoning coastal population and ever-increasing marine resource use bring to bear on the seas.

A metaphor concerning modern medicine is illustrative. As in marine ecology, scientific understanding of human physiology is incomplete, yet remedying health problems is imperative. Physicians are ultimately concerned with how well the body's systems are working and what causes underlying impairment. Structure, in this case anatomy, is only important in that it provides clues to function. So it is with the emerging field of marine conservation, where biologists work with systems ecologists

*Marine Protected Areas and Ocean Conservation,* by Tundi Spring Agardy.
© 1997 R.G. Landes Company.

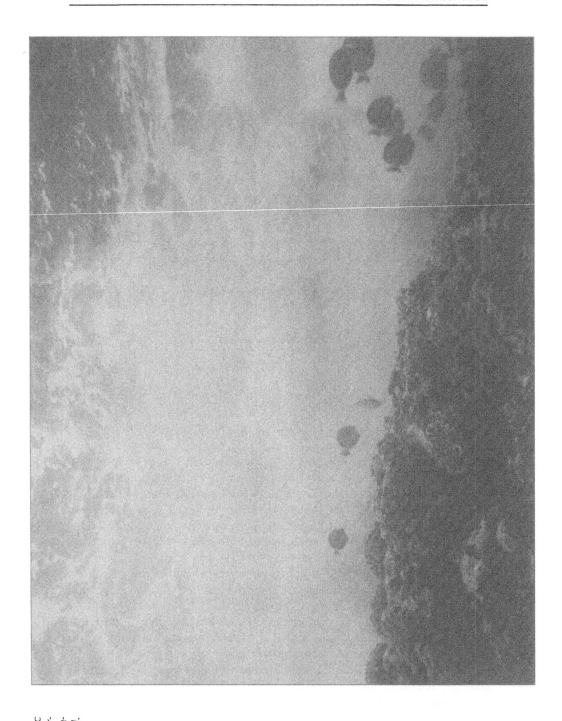

Fig. 4.1. The dynamic nature of life underwater: subsea Hawaii. Photo by J. Spring.

to elucidate how human impacts impair coastal and marine ecosystem function in order to know how to lessen those impacts. And multiple use areas and coastal management zones can provide the operating arena where such important work is carried out.

Process-directed conservation, where efforts are made to protect ecological processes and ecosystem functions rather than stocks of specific resources, is one of many important tools available for our utilization. Because we are constrained by both limited resources and limited time, some allocation of effort should go to targeting critical processes for conservation. Just as a physician targets the vital organs of his patient to maximize recovery, so this process-oriented conservation targets the ecological processes most vital to marine ecosystems.

Such a physiological approach to environmental protection must occur on two levels simultaneously. At the broad policy level, top-down approaches should help develop the science and policy of conservation to make it more effective and rational, much the same way that the medical profession sets broad policies and research agendas for improving collective human health.[1] On the more localized, field-based level, bottom-up approaches should aim to put conservation measures in place that protect the most vital processes, much the same way individual doctors promote the sustained health of individual patients by concentrating on the vital organs first and foremost. These two spheres of activity are not isolated, and the field projects play a critical role not only in dealing with conservation emergencies but also in demonstrating to policy-makers and funding institutions precisely how conservation derives benefits: advantages that can be couched in environmental, economic, social or ethical terms.[2]

The oceans and coasts harbor extremely complex ecosystems, with geographically widespread linkages and hierarchically-nested spatial and temporal scales.[3] The important ecological processes that create the conditions necessary for the maintenance of the ocean's great biodiversity and immense productivity are closely linked across these scales.[4] Despite this, we tend to want to model marine systems as if they were fully deterministic and conveniently simplistic, ignoring the realities of connectivity, complicated recruitment and population dynamics, and inherent chaos and uncertainty. As part of this desire to want to keep things simple, we cling to a mistaken idea that the acceptability of resource extraction or use can be evaluated by looking at the resource in question while ignoring all other parts of the system.

In order to effectively tackle the substantial marine conservation problems we face today, we need to ask clear questions about how we undermine ecosystem functions, how we can continue to use living resources sustainably, and how we can modify our behavior to ensure our survival and with it the rest of the planet's. And we must set out to derive those answers as quickly as we can.

Harnessing the science we currently have available to us to develop rigorous management should help ensure that use of marine resources and ocean space will be sustainable over long time frames. This science should draw from advances in the fields of ecology, population dynamics/genetics, physical oceanography and hydrology, and environmental studies. There are at least seven entry points for scientific methods in the conservation of marine biodiversity: 1) Defining the ecological bounds of the system and thus the appropriate geographical framework for management; 2) Identifying ecologically critical processes and areas; and allow relative ranking of an area's importance based on biodiversity or other criteria; 3) Assessing the <u>scientific</u> feasibility of a conservation or management project, including whether enough baseline information exists to develop ecologically-based management plans; 4) Defining management units for species of special concern, such as those that are threatened or endangered, have important ecological roles, high commercial value or are crucial to local culture, or act as indicator species; 5) Determining what levels of resource use can be sustained, using which technologies; 6) Highlighting the sectors in which integration of resource management is required (i.e., where the utilization of one resource will affect another); and 7) Monitoring to see if conservation objectives, both nature-centric and human-centric, are being met.

Scientifically-based, process-oriented conservation allows us to harness science to protect critical processes so that human communities can continue to rely on vital ecosystems. Safeguarding the processes which maintain complex ecosystems and their immense variety of life forms is difficult. It requires more than surveillance (what is there): it requires understanding (why it is there). In some instances, where components of the ecosystem are closely linked, it may be possible to "fence in" enough of the critical processes that structural and process-oriented conservation is achieved simultaneously. In most cases, however, a lack of basic understanding about critical linkages means that sooner or later, structure-oriented conservation will buckle under extrinsic development pressure and the indirect but insidious side effects of anthropogenic activity.

With all the visionary new concepts in conservation biology and resource management currently afloat, even the most conservative scientists agree that field-testing ideas is a prerequisite to embracing them (or, for that matter, tossing them blithely aside). Marine protected areas that are appropriate to the geographic scales of coastal and marine ecosystems, that contain management units grounded in ecology, and that allow multiple uses by establishing zoning to protect that which is most critical, most sensitive, or most amenable to monitoring and evaluation, can be the anchor to evaluate new ideas.[5,6] In many cases marine protected areas provide a unique opportunity to force definitions of vague concepts, field test them, evaluate their potential objectively, and demonstrate their usefulness. It can be seen as a unique opportunity precisely because our history of tinkering with the oceans is so far brief, and we haven't had the time yet to establish entrenched bureaucracies and rigid paradigms. It is an opportunity we must not waste.

Successful marine protected areas not only resolve local management issues but can also provide salient examples of how we should be managing our impacts on our seas in regional and even global scales. It is probably no exaggeration that the future of the earth's nearshore areas, to the extent that we have some role to play in deciding that future, rests firmly on the shoulders of the new generation of these protected areas.[7]

Is this a real departure from the status quo or merely new light shed on an old way of thinking? The flurry of recent papers on marine ecophysiology and on functionally-based approaches to conservation suggest the former (e.g., Lubchenco et al, 1991; Parsons, 1992).[8,9] Take, for instance, the scenario of protecting an estuary—those vital organs of the marine system that are so rich in ecological services and productivity. In the old days, a government agency charged with protecting such an estuary might have outlined the embayment on a map, fenced off its land margin, and posted signs alerting visitors and potential users of its protected status. Today the conservation effort would extend way beyond the boundaries of the bay itself by looking at critical linkages in nutrient cycling, migration of species, and systems links to other habitats in the watershed and out at sea.[10,11] And while the garrison reserve of yore would slowly get degraded by impacts from a distance, the functionally-based conservation scheme stands a chance of safeguarding that vital ecology for the future.

Arguably the most important role coastal and marine protected areas serve is as a starting point for exploring and delimiting functional linkages in coastal systems (Dayton, 1993), metaphorically moving us away from being quacks to being effective physicians.[12] Ecological studies that provide the basis for marine protected area work facilitate determination of appropriate boundaries for management units and specific framework for applying ecological principles for the purposes of management. Ecosystem management, seen by many as 'the joke at the party that everyone laughs at but no one gets', can be field tested in the context of protected areas. For those with a more terrestrial orientation ecosystem management implies using ecosystem science as a basis for management decisions that aim to maximize production of a commodity. For those with a marine background, ecosystem management typically means looking at the functional linkages between the target ecosystem and habitats or ecological communities outside in order to define functionally viable management units.[13] Both these interpretations of ecosystem management need a geographic context in which to be tested and coastal and marine protected areas provide the ideal venue for doing so.

Marine protected areas allow us to invoke the precautionary principle–a term that, like ecosystem management, has lost some of its
intrinsic value as it becomes popular political jargon without stringent
definition. In science-based conservation, the precautionary principle
is what drives managers to err on the side of conservatism when scientific uncertainty looms.[14,15] Central to the idea of the precautionary
principle is the notion that actions that produce irreversible change to
ecosystems (extinctions and the permanent restructuring of food webs,
for instance) must be avoided at all costs. Recognizing that the general status and condition of coastal and nearshore areas will undoubtedly decline and that scientific knowledge about marine and coastal
ecosystem functions is far from complete, marine protected areas provide a physical area in which to apply the precautionary principle and
buffer against unforeseen yet potentially disastrous management mistakes.

Coastal and marine protected areas also create on-the-ground (or
in-the-water!) frameworks for applying the idea of adaptive management. Adaptive management can be a nebulous term (Halbert, 1993;
Walters, 1987)—but fisheries science has provided us with some rigorous definitions of precisely what is meant by it (Holling, 1988).[16-18]
Two conditions must apply for resource management to be adaptive:
1) an explicit feedback loop between science and management must
be maintained so that management can be maximally flexible and responsive to both environmental and social changes; and 2) management measures must provide a setting for experimental manipulation
of regulations so that their efficacy can be objectively tested. As clearly-
recognizable entities marine protected areas can firmly establish such
management-science links and provide a laboratory for experimental
testing. This is all the more necessary in marine systems, where managers must deal with largely stochastic systems characterized by enormous
uncertainties.[19,20]

Some marine protected areas act as nodes in networks of monitoring sites designed to try and evaluate the general state of the marine
environment and specific conditions of nearshore ecosystems.[21] Such
monitoring and evaluation allows estimates of potential productivity
of renewable resources and is thus a major component in determining
sustainable levels of use. Marine protected area monitoring networks
also provide means to assess global change and field test theoretical
models of global scale processes.[22] Certain areas within reserves and
other protected areas, such as strictly protected core areas, also serve
as necessary controls against which the rate of environmental deterioration
can be gauged.[23]

As we gain more understanding of marine systems and their productivity, we reinforce intuitive beliefs that the management of our
impacts on ecological function must not be taken one-by-one. Cumulative impacts stretch over time and across space to collectively impair

function and undermine resilience. Establishing conservation measures that protect against the suite of anthropogenic impacts is notoriously difficult, and even the most idealistic among us recognize that triage is sometimes necessary. Since not all components of coastal and marine systems can be protected, human and financial resources should be targeted at those areas that harbor the most important ecological functions or those that are most threatened by direct and indirect human activity.[24] In this context, marine protected areas allow establishment of systems of the non-extractive zones or harvest refugia, in order to protect seed banks or sources of recruits and critical ecological processes that are currently being impaired or are likely to be impaired in the short term future.[25,26] There is increasing evidence that such refugia not only protect marine organisms in situ but that they can serve to increase productivity in a wider area.[27-29]

It may be doing conservation a disservice to separate the role that marine protected areas play in science-based management and the role they play in accommodating human needs, since the latter is a critical component of the former. Yet modern coastal and marine protected areas are so much more than laboratories for evaluating how scientifically rigorous our management measures are–they are often the only starting point for creating forums to resolve use conflicts and establish a basis for responsible use and responsible attitudes.[30] Marine protected areas in this context are publicly recognizable spaces which allow users to become actively involved in planning (rather than being the recipient of management regimes imposed from outside), and in management–including undertaking enforcement of regulations–through partnerships between regulatory agencies and user groups.[31,32] In this sense marine protected areas can provide a means to avert the tragedy of the commons and help foster a sense of stewardship for ocean resource and ocean space among the people who most rely on healthy, intact coastal systems.

Additionally, marine protected areas can act as a means to preserve traditional uses of resources or space that have remained sustainable over time. As eloquently stated by McNeely (1991), "local societies have ebbed and flowed as their wisdom was tested against the criterion of sustainability–those that were able to develop the wisdom, technology and knowledge to live within the limits of their environment were able to survive." By delimiting an area for the purpose of conservation, marine protected areas provide locales in which traditional and sustainable practices can continue to be undertaken by indigenous peoples.[33,34]

Establishing the new generation of marine protected areas is risk-laden. Frameworks for management in these marine protected areas must thus be sufficiently responsive and flexible to allow for change as better scientific information is gathered or conditions (environmental or social) change. Despite this, marine protected area planning must be done within the limits of a resource management community that is typically risk-averse. Getting administrators and government agencies to "buy in" to new models for marine conservation, especially those that recognize large scientific uncertainties, and put more of management in the hands of the users, requires patience and compromise. At the same time, we stand at a critical juncture with respect to the future of marine biological diversity and ocean health.[35] We can't afford to be patient and plodding much longer.

It must at the same time be emphasized that the survival and efficacy of marine protected areas is inevitably linked to the larger matrix in which they are planned and carried out. No marine protected area is an island; the extensive linkages and amorphous nature of boundaries make context all the more important. Sadly, if we allow the world outside marine protected areas to continue to decline in response to myriad, chronic impacts, even the most well-designed and executed protected areas have no future. Nonetheless, marine protected areas serve as valuable anchors for the large scale conservation of the biosphere, and as such they secure the future of marine conservation.

## SECTION 2. MANAGEMENT GOALS AND OBJECTIVES OF MARINE PROTECTED AREAS

### 2.1 GENERIC OBJECTIVES

The ultimate goal of any marine protected area is marine conservation–that is, the protection of critical ecological processes that maintain the ecosystem and allow for the production of goods and services beneficial to humankind, while allowing for utilization of ocean space and resources that is sustainable in an ecological sense. However, there are as many specific goals or objectives for marine protected area establishment as there are existing marine protected areas. Much of the scientific and sociological literature pertaining to marine protected areas and ocean conservation lists objectives of various types of protected areas (see especially Salm and Clark, 1984, and, more recently, Kelleher and Kenchington, 1994, as well as Jones, 1994).[6,30,36] The latter paper lists the following, organized according to whether the objectives are scientific, economic of cultural/ethical: 1) maintain genetic/species diversity; 2) promote research; 3) allow creation of education and training areas; 4) conserve habitat and biota; 5) allow for baseline monitoring; 6) protect rare/important species (all scientific goals); 7) promote tourism and recreation; 8) promote sustainable development; 9) recolonize exploited areas; 10) protect coastlines; 11) allow for alternative economic development (all economic goals); 12) preserve aesthetic value; 13) protect historic/cultural sites; 14) exert political influence or assert jurisdiction; and 15) protect intrinsic and/or absolute value of an area (cultural and ethical concerns).

## 2.2. GOALS FOR MARINE PROTECTED AREA ESTABLISHMENT

The myriad objectives listed above may be organized into seven broad goals, all having to do with how humans value marine resources and what obstacles there are to effective management of marine resources. First, marine protected areas can help to overcome the "out of sight, out of mind" phenomenon that plagues would-be stewards of marine resources, by providing a sense of place that people can relate to and in which they can take ownership.[37] By delimiting a clearly-defined, concrete and manageably-sized area, protected area planners and managers can focus attention, concern, and management resources on a particular site.

An second goal for marine protected area establishment that is getting much recent attention is to provide a testing ground for management.

*Fig. 4.2. Ecotourism: whale-watching off Maui, Hawaii. Photo by J. Spring.*

Such a testing ground will provide answers to questions that commonly arise about marine conservation, e.g., "Can resource use in this area can managed feasibly?" and "How can management of this coastal or marine area be made to be as integrated as possible?" If management of ocean space and marine and coastal use can be undertaken efficiently and with maximum benefit to users in a marine protected area, such management could be expanded to include greater areas, including state or provincial coastal zones or even entire coastal zones of a nation.

The third general type of goal for marine protected area establishment has to do with the potential social benefits such areas provide local communities. The creation of a new jurisdictional entity in the coastal area can often act to empower local users who might not otherwise have a collective voice in decision-making about resource use and allocation. Furthermore, marine protected areas that involve local communities in planning and implementation often allow for more equitable sharing of benefits than might have existed previously.

Marine protected areas also allow us to gain better information about marine ecology and human impacts on it. As mentioned repeatedly in this book and others on the subject of the oceans, we are hindered from effective management of marine resources and space by our ignorance. The gaps in information about marine ecosystems are enormous, particularly in comparison to what we know about terrestrial ecosystems. Marine protected areas provide the sole means for establishing control areas and sites for experimental manipulation—without which our knowledge will remain riddled with gaps. Those marine protected areas that allow managers to set up true systems of adaptive management, whereby management regimes can be tested for their efficacy, are especially valuable in this regard.

The fourth broad goal—that dealing with protected areas as a tool to regulate levels of natural resource harvest—is very broad indeed. Most marine protected areas established to date have had this goal in mind: to enable development, of an area or a resource (or set of resources) to be undertaken in a sustainable fashion. Protected areas used in the management of fisheries, for example, allow limits to be set on the size, number, time of take, and/or method of harvest to ensure that fish stocks will be conserved. On the other end of the spectrum of protected areas by size and ambitiousness, multiple use coastal planning areas allow sensitive or ecologically valuable areas to be preserved while regulated use is allowed in other areas.

This raises the question of species of special concern and how they might be protected through reserves, sanctuaries, or other designations. These species are ones that are highly valued by the public, and which can be used for raising awareness about marine issues more broadly. More and more, marine protected areas are being used to conserve species of special concern such as sea turtles, whales, endangered sharks, etc., or especially sensitive habitats such as seagrass beds or forests of rare branching corals. Such special species and habitats have unique ecological requirements, thus the protected areas will be designed with these in mind.[38]

The last broad goal is the most tenuous and vague—but it is increasingly being invoked as the reason for designating coastal and marine protected sites. Such areas are protected as buffers against unforeseeable future management mistakes—allowing managers to put the precautionary principle into practice.[2] Systems of representative protected areas, for instance, allow nations to conserve at least one type of ecosystem or habitat in perpetuity, so that even if uncontrolled development alters or destroys other similar areas, representative areas will be left intact.[39] Protected areas that are established with this objective in mind allow us to err on the side on conservation instead of on the side of irreversible extinction or degradation.

## SECTION 3. STRATEGIES FOR DESIGNATING NETWORKS OF MPAs

### 3.1 WHY MARINE PROTECTED AREA NETWORKS ARE NEEDED

Marine protected areas are essentially islands: islands of controlled and sustainable use and conservation of biodiversity surrounded by a sea of mismanagement, overexploitation, and open access. And as such, marine protected areas will not promote marine conservation more widely unless they are designated in a systematic way that takes into account the entire ecosystems of which they are a part, and in fact the planetary context itself. Like other more familiar protected areas like national parks and other terrestrial reserves, marine protected areas will be rendered ineffective if the seas around them are degraded, if they are the last refuge for species, and if large scale processes that underlie such degradation are not tackled in regional or global contexts.

The "island effect", as it is known, is theoretically derived from MacArthur and Wilson's classic island biogeography theory.[40] Based on this theory and more recent ecological modelling, it is clear that there are many reasons why biota that are confined to islands of protection (which are effectively the same as true islands in a biogeographic sense) are inherently vulnerable; these include, inter alia, 1) population sizes are suppressed; 2) migration may be curtailed leading to changes in behavior, bioenergetics, etc.; 3) genetic heterogeneity is potentially reduced due to increased opportunities for inbreeding; and 4) habitats and the ecological processes they support are at risk from external impacts along the edge of protected areas. Thus populations of organisms and sometimes species themselves that live within the bounds of garrison reserves, whether on land or in the sea, stand threatened—as do the ecological processes that support the biodiversity we seek to conserve.

For these reasons. it is best to think of marine protected areas as having, in addition to whatever benefits are afforded through in situ conservation, benefits in acting as single nodes within a framework of networks that act to: a) counter some of the threats facing single protected areas; and b) lay the groundwork for national, regional, and global policies that prevent further degradation of the seas around them. These networks should be the long term goals for coastal and marine conservation that utilizes marine protected areas, even if political, economic, and other constraints limit the immediate implementation of entire networks.

## 3.2 STRATEGIES FOR DEVELOPING MARINE PROTECTED AREA NETWORKS

The interconnectedness of habitats and ecological processes in marine and coastal environments means that effective conservation will have to target not only a particular place (of importance to either man, an ecosystem, or the biosphere) but the larger context in which that area of interest sits. A regional approach to designating multiple marine protected areas is thus required for comprehensive and holistic ocean conservation.[41] A regional approach has many benefits: it allows development of the big picture view, it relies on a functional, ecosystem-based outlook, and it identifies and paves the way for opportunities to undertake co-management, joint management of shared resources, and regional agreements to protect whole ecosystems.

Given the various reasons that decision-makers or user groups might want to establish protected areas, are there strategic means to designing networks of marine protected areas to counter real and prospective threats to marine systems? The answer is yes. Strategies for establishing networks of marine and coastal protected areas fall under three approaches: 1) preservation of ocean or coastal "wilderness" areas (I use the term wilderness in quotation marks because no part of the world's oceans, inland seas, or coastlines is pristine); 2) resolution of conflicts among users (current or in the future); or 3) restoration of degraded or overexploited areas.

Most existing national marine protected area networks follow the first strategy.[42] For instance, Parks Canada is currently designing a network of Marine National Conservation Areas to represent each of the 29 distinct ecoregions of Canada's Atlantic, Great Lakes, Pacific, and Arctic coasts. The long term goal of this programme is to establish a protected area in each region.[43] Similarly, the federal government of Australia is developing a strategy for establishing a National Representative System within Australian Coastal and Marine Environments (P. Burbridge, Director, Australian National Parks and Wildlife Service, personal communication). In designing such a system, site selection will be guided by representativeness, opportunity, and redundancy (meaning that the government's policy is to designate more than one protected area per representative habitat type). Other national efforts are currently underway.[44,45] In fact, the 1995 publication of the Great Barrier Reef Marine Park Authority, the World Bank, and IUCN (Kelleher, et al, 1995) which is the most comprehensive overview of existing marine protected areas and gaps in coverage, strongly urges all countries to establish such representative networks.[44]

Conflict resolution is the other major driving force behind the establishment of networks or systems of reserves or protected areas. Virtually all the world's coasts and nearshore areas are characterized by conflict between user groups or jurisdictional agencies. Shipping and mineral extraction, for instance, often conflict with recreational use of coastal areas. Fishing, both commercial and subsistence, conflicts with skin and scuba diving and nature-based tourism. But other less obvious conflicts also occur. The use of one fishing gear type may directly conflict with the use of another. State or provincial government sanctions on the use of an area may conflict with national security or economic concerns. Visitors to a coastal area may offend local communities having different cultural norms and beliefs. Marine protected areas allow much of this conflict to be minimized or resolved through segregation of uses and stakeholder participation in the process. Noteworthy examples of such protected area systems include the U.S. Marine Sanctuary Program and the emergent Arctic Protected Areas plan.

Finally, a third approach to marine protected area designation is to look at threats to ecosystems and degree of degradation of areas, and establish a system of marine protected areas to allow restoration of sites (and replenishment of resources) as quickly as possible. Though few systematic attempts to identify coastal and marine areas in need of restoration exist, the ongoing restoration program for South Florida (USA) (including the Everglades area, Florida Bay, and the Florida Keys) is a good example of an analytical approach to establishing a network of protected areas for restoration purposes.[46]

## REFERENCES

1. Bensted-Smith, R. and S. Cobb. Reform of protected area institutions in East Africa. Parks 1996;5(3): 3-19.
2. Gubbay, S. Marine Protected Areas: Principles and Techniques for Management. London, Chapman and Hall, 1995.
3. O'Neill, R.V., D.L. DeAngelis, J.B. Waide, and T.F.H. Allen. A Hierarchical Concept of Ecosystems. Monographs in Population Biology 1986; 23. Princeton NJ, Princeton University Press.
4. National Academy of Sciences. Understanding Marine Biodiversity. Washington, DC, National Academy Press, 1995.
5. Bjorklund, M.I. Achievements in marine conservation: International marine parks. Env. Cons. 1974;1(3): 205-223.
6. Salm, R.V. and J.A. Dobbin. Management and administration of marine protected areas. In: Proc. of the Workshop in the Application of the Bio-

sphere Reserve Concept to Coastal Areas, 14-20 August 1989, San Francisco, CA, USA.

7. Agardy, T. The Science of Conservation in the Coastal Zone: New Insights on How to Design, Implement and Monitor Marine Protected Areas. Proc. of the World Parks Congress 8-21 Feb. 1992, Caracas, Venezuela. IUCN, Gland, Switzerland. 1995.

8. Lubchenco, J. et al. The Sustainable Biosphere Initiative: an ecological research agenda. Ecology 1991; 72: 371-412.

9. Parsons, T.R. Biological coastal communities: productivity and impacts. In: Coastal Systems Studies and Sustainable Development. Proceedings of the COMAR Interregional Scientific Conference, UNESCO, Paris, 21-25 May 1991. UNESCO Reports in Marine Science 1992; 64:27-37.

10. Bakun, A. Definition of environmental variability affecting biological processes in large marine ecosystems. In: K. Sherman and L. Herander, eds. Variability and Management of Large Marine Ecosystems, Washington, DC AAAS Press, 1986: 89-107.

11. Costanza, R., W.M. Kemp and W.R. Boynton. Predictability, scale, and biodiversity in coastal and estuarine ecosystems: implications for management. Ambio 1993; 22(2-3):88-96.

12. Dayton, P.K. Scaling, disturbance, and dynamics: Stability of benthic marine communities. In: T. Agardy, ed. The Science of Conservation in the Coastal Zone. Proceedings of the IVth World Conference on Parks and Protected Areas. 1993 IUCN Marine Conservation & Development Report, Gland, Switzerland.

13. Kenchington, R.A. and M.T. Agardy. Achieving marine conservation through biosphere reserve planning. Env. Cons. 1990; 17(1): 39-44.

14. Ludwig, D., R. Hilborn, and C. Walters. Uncertainty, resource exploitation, and conservation: lessons from history. Science 1993; 260(2): 17;36.

15. Slocombe, D.S. Implementing ecosystem-based management. Bioscience 1993; 43(9):612-622.

16. Halbert, C.L. How adaptive is adaptive management? Implementing adaptive management in Washington State and British Columbia. Reviews in Fisheries Science 1993;1(3):261-283.

17. Walters, C. Adaptive Management of Renewable Resources. New York, Macmillan Publ., 1987.

18. Holling, C.S. Adaptive Environmental Assessment and Management. New York, John Wiley and Sons, 1988.

19. Mann, K.H. Physical oceanography, food chains, and fish stocks: a review. ICES J. Mar. Sci. 1993; 50:105-119.

20. Mann, K.H. and J.R. Lazier. Dynamics of Marine Ecosystems. Oxford, Blackwell Scientific Publications, 1991.

21. International Union for the Conservation of Nature and Natural Resources (IUCN). Parks for Life: Report of the IVth World Parks Congress on National Parks and Protected Areas. IUCN, Gland, Switzerland, 1993.

22. Agardy, T. Coral reefs and mangrove systems as bio-indicators of large scale phenomena: a perspective on climate change. In: Proc. of the Symp.

on Biol. Indicators of Global Change 7-9 May 1992 Brussels, Belgium. J.J. Symoens, P. Devos, J. Rammeloo and C. Verstraeten, eds. Royal Academy of Overseas Sciences, 1994:93-105.

23. Yurick, D.B. International networking of marine sanctuaries. Oceanus 1988; 31(1): 82-87.

24. Agardy, T. Advances in marine conservation: the role of protected areas. Trends in Ecology and Evolution 1994; 9(7):2676-270.

25. Ballantine, W.J. Marine reserves for New Zealand. Univ. of Auckland, Leigh Lab. Bull., 1991;25.

26. Bohnsack, J.A. The potential of marine fishery reserves for reef fish management in the U.S. Southern Atlantic. NOAA Tech. Mem/ NMFS-SEFC-261, 1990.

27. Alcala, A.C. Effects of marine reserves on coral fish abundances and yields of Philippine coral reefs. Ambio 1988; 17:194-199.

28. Alcala, A.C. and G.R. Russ. A direct test of the effects of protective management on abundance and yield of tropical marine resources. Journal du Conseil 1990;47(1): 40-47.

29. Ballantine, W.J. Networks of "no-take" marine reserves are practical and necessary. In: Marine Protected Areas and Sustainable Fisheries. N.L. Shackell and J.H.M. Willison, eds. Science and Management of Protected Areas Association, Wolfville, Nova Scotia, 1995: 13-20.

30. Kelleher, G.B. and R.A. Kenchington. Guidelines for Establishing Marine Protected Areas. IUCN Marine Conservation and Development Report, Gland, Switzerland, 1992.

31. Smith, A.H. and F. Berkes. Solutions to the 'tragedy of the commons': sea-urchin management in St. Lucia, West Indes. Env. Cons. 1991; 18(2):131-136.

32. White, A.T. and V. P. Palaganas. Philippine Tubbataha Reef National Marine Park: status, management issues, and proposed plan. Env. Cons. 1991; 18(2):148-157;136.

# PART III:
# TYPES OF MARINE
# PROTECTED AREAS

# CHAPTER 5

# MARINE PROTECTED AREA TYPOLOGIES

## SECTION 1. MARINE MANAGEMENT AREAS AND MPAs

In the broadest sense, marine management areas can be defined as any area of the coastal zone or open ocean conferred some level of protection for the purpose of managing use of resources and ocean space or protecting vulnerable or threatened habitats and species. Thus marine management areas (MMAs) can be thought of as any coastal or marine area that is managed for conservation purposes (inter alia), and marine protected areas as we traditionally think of them would be one subset of MMA categories. Like all classification schemes, this MMA typology is artificial in the sense that many types of managed areas cannot be clearly defined as belonging to one class or another, and some marine protected areas seem to fit more than one management category. Nonetheless the classification is useful in that it demonstrates the wide spectrum of management regimes that are used to control our use of the coasts and oceans.

A minimum of least seven major categories of marine management areas can be distinguished today. These major classes of MMAs are described in detail below. The classification scheme begins with the most narrow and spatially limited type of Marine Management Area and extends to the broadest and largest category recognized.

### CATEGORY 1: CLOSED AREAS

Closed areas include fisheries harvest refugia, moratorium or areas closed to OCS activity, and similar areas where a certain class of use is restricted for the main purpose of ensuring the sustainability of resources. Closed areas differ from Sensitive Sea Areas (Category 3) in that management of a specific type of use is the main objective for establishing the moratorium. Closed areas can be and often are temporary or seasonal.

---

## CATEGORY 2: RESEARCH AND MONITORING AREAS

These marine management areas are designated either as controls in experimental science or sites for monitoring to evaluate environmental condition, or are protected as natural laboratories to support basic research in ecology, fisheries, oceanography, etc. Research sites are managed specifically to control variables or to allow for inter-site comparisons, and can either exist as independent entities (for example, long-term ecological research (LTER) sites) or as core areas within multiple use reserves.

## CATEGORY 3: SENSITIVE SEA AREAS

The International Maritime Organization (IMO) recognizes Sensitive Sea Areas as areas that need special protection through action by the IMO because of their ecological or socioeconomic significance and their vulnerability to damage by maritime activities. Such areas include coral reef areas or temperate sounds where ship transit is prohibited for reasons of safety and environmental sensitivity.

## CATEGORY 4: MARINE SANCTUARIES AND MARINE PARKS

Marine protected areas such as marine sanctuaries and traditional marine parks constitute a broad and complex assemblage of marine management areas. The World Conservation Union (IUCN) previously recognized ten classes of marine protected area, including: strict nature reserve, national park, natural monument, wildlife sanctuary, protected landscape, resource reserve, natural biotic area or anthropological reserve, multiple use management area, biosphere reserve, and world heritage site. These have been recently (IUCN 1994) narrowed down to only six categories for both terrestrial and marine protected areas (see below).[1] Clearly some of these categories overlap with the seven main categories described here. Nonetheless, the feature common to all marine parks and sanctuaries is the fact that they are established to accommodate a set of particular uses while conserving the coastal or marine ecosystem and its processes. In this taxonomy, marine parks and sanctuaries range from the seaward extensions of coastal terrestrial parks, to ecosystem-based multiple use marine parks, sanctuaries and biosphere reserves.

## CATEGORY 5: REGIONAL SEAS AND LARGE MARINE ECOSYSTEM AREAS

Regional Seas are formally recognized by the United Nations Environment Programme as enclosed or semi-enclosed seas that fall under the jurisdiction of more than one nation. Regional Seas become marine managed areas when bilateral or multilateral agreements are drawn up to control pollution, develop cooperatively managed areas (e.g. transboundary reserves), and allow for joint management of endangered species or commercially-important renewable resources. Large marine ecosystems are areas that represent a coherent ecological unit (whether enclosed or semi-enclosed seas or biogeographic ocean areas).[2] There are currently forty-nine delineated Large Marine Ecosystems recognized by marine ecologists worldwide (Fig. 1.1). These LMEs sometimes form the basis for Regional Seas agreements (see also chapter 11 for a discussion of Regional Seas treaty agreements).

### CATEGORY 6: INTEGRATED MANAGEMENT AREAS
Integrated management areas include state-administered coastal zone planning areas and Exclusive Economic Zones managed by federal authorities. Management of state or provincial coastal zone areas tends to be more coordinated and integrated because it falls under the purview of a single management authority in each jurisdiction, whereas federally-managed EEZs fall under the responsibility of many agencies.

### CATEGORY 7: HIGH SEAS UNDER THE U.N. LAW OF THE SEA TREATY
Although the high seas technically constitute a global commons and are therefore not a managed marine area, international treaties and codified customary law do create a cooperative management regime for those states that sign and ratify these agreements. The UN Law of the Sea Treaty has now been ratified by most coastal nations (see the chapter 11 section on UNCLOS), signifying a desire to standardize jurisdiction, management, and conservation of coastal and high seas areas around the globe.

## SECTION 2. IUCN CLASSIFICATION SCHEME
The World Conservation Union (IUCN) has been in the business of categorizing protected areas since the term "national park" was first defined at its 1969 General Assembly.[1] In 1978, the IUCN Commission on National Parks and Protected Areas (CNPPA) produced a report proposing ten categories of protected areas (marine and terrestrial protected areas are not dealt with separately). However, the use of these categories highlighted some difficulties with them, and in 1994 IUCN put forth a revised categorization. The six currently accepted protected area designations apply equally well for marine or terrestrial settings. They include:

### CATEGORY I A: STRICT NATURE RESERVE
The area is defined as "an area of land and/or sea possessing some outstanding or representative ecosystems, geological or physiological features and/or species, available primarily for scientific research and/or environmental monitoring." In the marine setting, closed areas or coastal refuge areas would fit this description.

## CATEGORY I B: WILDERNESS AREA

This protected area type is defined as "a large area of unmodified or slightly modified land, and/or sea, retaining its natural character and influence, without permanent or significant habitation, which is protected and managed so as to preserve its natural condition." No open ocean wilderness areas exist at this time.

## CATEGORY II: NATIONAL PARK

This is defined as a "natural area of land and/or sea, designated to (a) protect the ecological integrity of one or more ecosystems for present and future generations; (b) exclude exploitation or occupation inimical to the purposes of the designation of the area; and (c) provide a foundation for spiritual, scientific, educational, recreational and visitor opportunities, all of which must be environmentally and culturally compatible." Marine National Parks fit this description.

## CATEGORY III: NATURAL MONUMENT

These are "areas containing one or more specific natural or natural/cultural feature which is outstanding or unique in value because of its inherent rarity, representative or aesthetic qualities or cultural significance." Marine National Monuments would fall into this category, however a paucity of such areas currently exist.

## CATEGORY IV: HABITAT/SPECIES MANAGEMENT AREA

This category covers "areas of land and/or sea subject to active intervention for management purposes so as to ensure the maintenance of habitats and/or to meet the requirements of specific species." In the marine realm, such areas would include Nature Conservation Areas, Wildlife Sanctuaries (e.g., coastal bird rookeries, the Southern Ocean Whale Sanctuary) and some managed areas for fisheries.

## CATEGORY V: PROTECTED LANDSCAPE/SEASCAPE

This is defined as an "area of land, with coasts and sea as appropriate, where the interaction of people and nature over time has produced an area of distinct character with significant aesthetic, ecological and/or cultural value, and often with high biological diversity." Some coastal biosphere reserves would likely fit this category.

## CATEGORY VI: MANAGED RESOURCE PROTECTED AREA

This broadest category of IUCN-defined protected areas is defined as an "area containing predominantly unmodified natural systems, managed to ensure long-term protection and maintenance of biological diversity, while providing at the same time a sustainable flow of natural products and services to meet community needs." Multiple use marine protected areas, including biosphere reserves, would all likely fit this latter category.

## SECTION 3. FUNCTIONAL CLASSIFICATION OF MPAs

Today's marine protected areas represent a decided departure from the rigid, limited management tools of the past and the umbilical cord link to terrestrial park planning. Management areas are no longer established as amusement parks for recreational use or as shaded areas on a map with little in the way of regulation behind them. Coastal planning and marine protected area management are becoming sophisticated initiatives employing new methods and tools. The new generation of marine protected areas are now largely represented by multiple use reserves accommodating many different users, each with their own objectives. Administrators are finding different uses can indeed be fostered without adverse impacts on ecosystem function, as long as planning is based on ecological realities and relies on specific objectives from the outset.

Marine protected areas serve a great range of conservation and management objectives and exist in a wide array of designs. It is important to recognize that today's marine protected areas are not merely oceanic recreational areas, nor are they exclusively extensions of coastal, but terrestrially-focused, parks—even though this is still the way the public largely views them. Modern marine protected areas range from small areas designated to maximize an area's value to a single set of stakeholders to very large, multiple use areas meant to accommodate a wide variety of user groups. The common thread that ties these disparate types of protected areas together is that the implementation of the protected area provides a concrete framework for developing, executing or testing management measures. In essence, a marine protected area allows stakeholders—whether user groups, scientific researchers, resource managers, or government agencies—to flag an area as worthy of interest and protection. Such protected areas are thus a footing for integrated management and better governance.

Arguments abound about the nature of marine protected areas and how they relate to conventional land parks. Whether the differences between terrestrial and marine protected areas have to do with differences in kind or degree is irrelevant; the fact remains that marine protected areas do significantly differ from protected areas on land. The greatest single factor underlying this difference is the nebulous nature of boundaries in the fluid environment of the sea. It is notoriously difficult to attach boundary conditions to marine ecological processes, just as it is difficult to bound the impacts that affect those processes. In essence, it is impossible to 'fence in' living marine resources or the critical ecological processes that support them, just as it is impossible to 'fence out' the degradation of ocean environments caused by land-based sources of pollution, changes in hydrology, or ecological disruptions occurring in areas adjacent or linked to a protected area. And this holds true not only for open ocean pelagic environments but for the coastal zones as well, where functional linkages between habitats are so geographically widespread.

The open nature of coastal and ocean areas exists as a spectrum ranging from relatively fixed and "land-like" systems to highly dynamic and complex systems. Coral reef ecosystems, for instance, harbor organisms that are largely confined in their movements to the specific habitats of reef, surrounding soft or hard benthos, and coastal wetlands.[3] The structural framework for reef systems is fixed in place and can be mapped, much like a tropical forest provides a relatively fixed framework for the interactions of the forest community. The functional links between the water column in reefal areas and the benthos are strong so one can treat the ocean space together with reef structures themselves. In contrast, temperate open ocean systems such as estuarine/gulf/banks complexes are highly dynamic and in no way "fixed." Here living marine resources move in space and time according to physically-dominated, largely non-deterministic patterns. The ecology of the benthos is not strongly linked to that of the water column, and physical reference points for the system cannot easily be mapped.

The wide array of system types thus presents a challenge to conservationists and resource managers, requiring that protected area measures be appropriate to the system in question. The random application of terrestrial models to marine environments will not result in a viable means of protecting resources and the underlying ecology that gives rise to them. New paradigms are needed—and the newest generation of marine protected areas reflects this new way of thinking.

Modern marine protected areas serve a wide variety of functions. The size, shape and means of implementation of any protected area is a function of the primary objectives it has been targeted to achieve. If the objective of a protected area is, for instance, the protection of a single vulnerable habitat type from a specific type of use (e.g., protection of a fringing reef system from prospective shipping accidents), the resulting protected area can be simple in both design and management. If, however, the objectives are varied and target a wide range of habitats/resources, the protected area will have to be necessarily more complex. In most cases worldwide, the complexity of marine conservation problems requiring solution is such that only large, multiple use protected areas can provide a viable framework for conservation.

Modern, effective marine protected areas are characterized by a number of common attributes. Where a functional approach is adhered to, in other words where the object of conservation is not a single stock of resources or a single species but the ecosystem and its processes, marine protected areas will tend to be large and encompass many types of linked habitats. These large multiple use protected areas can be thought of as demonstrating the concept of ecosystem-based management, where the limits of protection in a geographical sense are based on the extent to which movements of organisms and physically-linked processes. The underlying ecology thus defines the outer boundaries for the area of protection. In recognizing these linkages, marine protected area planners can work towards conserving ecosystem integrity, not just individual resources or ecosystem structure.

That modern marine protected areas are designed with ecological linkages in mind does not mean that all are similar in plan or execution, even when similar habitat types are being protected. This is because the ultimate nature of the protected area reflects the specific objectives for which it was created. The objective-driven nature of multiple use marine protected areas is most clearly demonstrated in the type of zoning exhibited in the protected area plan. Zoning is used to accommodate a wide variety of user groups in relative harmony, and can be a tool for dispute resolution where conflicting uses clash.

According to a functional view of marine protected area categorization, there are four types of marine protected areas. The first type is perhaps most common worldwide, and exists as extensions of terrestrial parks. Typically these parks are formed when management objectives are broadened to include conservation of coastal areas. For example, the terrestrial Virgin Islands National Park in St. John, U.S. Virgin Islands, had until recently little influence on activities in near shore waters adjacent to the coastal park. However, in 1981 the Virgin Islands National Parks became a coastal Biosphere Reserve under the United Nations Educational, Scientific, and Cultural Organization's Man and Biosphere Program.[4] The expansion of the area of interest allowed local communities formerly living outside the park to participate more fully in management, and allowed the U.S. Park Service to exert more control over marine activities that were impacting the coastal ecosystem.

The second type of marine protected area is the small scale reserve or park established to reach a single objective. Saba Island in the Netherlands Antilles in one such example.[5] Saba Island is a small extinct volcano that rises steeply from the deep ocean floor of the Lesser Antillean chain. Despite its small size (9 km² in area) and population (approximately 1000 inhabitants), it has succeeded in attracting and maintaining a thriving tourist trade. The main reason people come to Saba Island is to scuba dive: the volcanic cone plunges dramatically from mountainside to seafloor, creating beautiful undersea walls and pinnacles. Fish are abundant; the productive Saba Banks are situated to the west and support large stocks of pelagics, and the fringing reefs around the island harbor many reef fish species. The general public now knows that the diversity of coral reefs has been shown to rival that of rain forests (Norse, 1993), and the diverse marine biota of Saba' environs are a lure for ecotourists and divers alike.[6] The coastal zone is relatively unperturbed, since waterfront development is close to impossible on the steep slopes of the island, agriculture and industry are practiced only in the small scale, and population size remains low.

The Saba Marine Park encompasses the entire coastal zone of the island, from mean high water to a depth of approximately 60 meters, and two offshore sea mounts (Fig. 5.1). The Park was established in 1987, as a means of controlling the activities of dive boat and charter boat operators in Saba's waters. The creation of the Park was very much a proactive, forward-thinking measure, since environmental quality remains high. Nonetheless, the residents of Saba felt it best if anchor damage and the take of marine organisms was minimized through the establishment of a multiple use protected area.

The Park is divided into four zones: 1) multiple use zones in which fishing and diving are permitted; 2) recreational diving zones where fishing and anchoring are not permitted; 3) anchor zones in which free anchoring and mooring are permitted; and 4) recreational zones which allow swimming, boating, snorkeling, diving and fishing.[5] Users' conflicts are largely avoided, impacts on the environment are curtailed, and public education about ecological processes is made available through park leaflets, brochures, and signs. Since the creation of the Park, tourism has increased, in part due to the attractiveness of pristine protected areas to ecotourists. User fees generated by the "dollar-a-dive" rule allow Sabans to generate enough revenues to support patrols, buoy maintenance, and visitor services.[7] Sabans are proud of their Park, and do much to promote the marketing potential of this tourist destination.

At the other end of the size spectrum from small single objective marine parks is the large coastal and marine protected area zoned for multiple uses. Since coastal and marine areas the world over provide food, transportation, recreation, and energy resources to increasing numbers of people each year and demands for these resources are rising, the potential for user conflicts is radically heightened. Traditional uses of coastal resources are often displaced by profitable but non-conservative technologies which preclude effective, comprehensive, and long-term management. This situation can be avoided or counteracted by instigating proactive multiple use planning in which all users can be accommodated in a sustainable way. However, such multiple use zoning plans can only exist in a concrete management framework: marine and coastal protected areas provide just such a foundation.

Oceans and coastal areas are many things to many people. To commercial and artisanal fishermen and their customers they are a seemingly limitless breadbasket there for the taking. For anglers, yachtsmen, surfers, swimmers, etc., they present boundless opportunities for recreation. To energy and shipping technologies the seas present an invaluable industrial resource. For some, the shoreline and oceans have some unquantifiable yet important spiritual value. These human perceptions of the importance of marine areas exist exclusive of the fact that the oceans and coastal margins play a vital role in maintaining

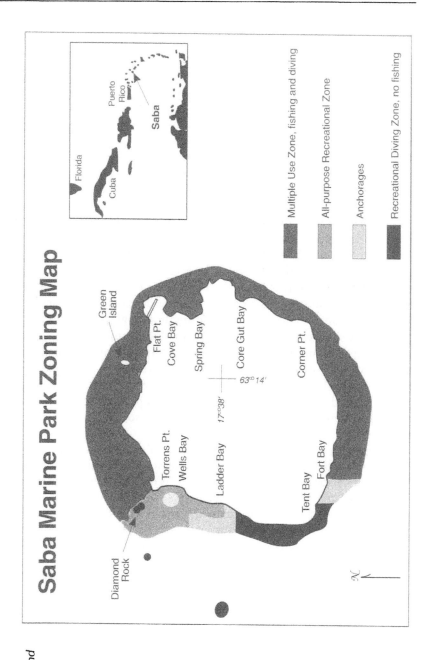

Fig. 5.1. The Saba Island
Marine Park.

the biosphere itself.[8] The value of the coastal zone is thus more than the sum of its calculable parts.

Despite this, the coastal zones remain in peril—thanks in large part to conflicts over resource use. No overseeing steward for ocean and coastal areas exists: in many countries management of marine areas, if it exists at all, is an auxiliary function in the naval or commerce branches of government. Hardin's (1966) "tragedy of the commons" has been exemplified time and time again.[9] And all sectors continue to increase their demands on ocean space and ocean resources, even as they continue to be degraded.

One logical way to accommodate all users and avoid resource conflict in areas where coastal populations are on the rise is to undertake comprehensive and proactive planning.[10,11] This planning commonly utilizes zoning for different uses and degrees of use, within the context of a marine park or other protected area. Multiple use marine and coastal areas were originally developed in response to increasing pressures and conflicts in the coastal zones of developed countries, largely as an extension of land use planning methodologies being used in the United States, Europe, and Australia. In the last decade, however, multiple use planning in marine areas has diverged from land use planning in where and how such areas are planned. Increasingly, multiple use areas are established as community-based protected areas in developing countries, and in both developing and developed country applications ecological considerations, rather than purely social ones, provide the geographical framework for zoning.[12,13]

Some areas, notably the Great Barrier Reef Marine Park in Australia, use such multiple use zoning schemes to minimize impacts on ecologically critical or sensitive areas.[14,15] When these multiple use protected areas are based on good scientific information regarding ecological processes and critical areas, the needs of local inhabitants, visiting tourists, scientists, and industry can be met in a system that meets minimum requirements for keeping ecosystems intact and productive. Unfortunately, examples like the Great Barrier Reef Marine Park that have been in place long enough to demonstrate concrete social and ecological benefits are rare.

Most successful multiple use marine or coastal protected areas aim to conserve critical ecological processes as well as threatened habitats and species. At times this is a corollary benefit, if the main intent of the marine protected area establishment is to protect human interests directly. Marine protected areas can be established to empower local communities, vesting them with responsibility for use of ocean space or resources, or guaranteeing access. The various independent objectives that marine protected areas may target are given in Table 5.1.

Successful multiple use protected areas will target many of these objectives simultaneously, but only in ways that are feasible and cost-effective.

---

### Table 5.1. Objectives of multiple use marine protected areas.

1) to safeguard traditional sustainable uses;
2) to serve as centers for public education and schooling;
3) to act as models for training programs in coastal zone management;
4) to serve as research stations for monitoring and experimental research;
5) to provide controlled habitats for ecological restoration;
6) to guarantee public access to shorelines;
7) to institute a means to limit entry to an area or to a particular user group;
8) to facilitate the political empowerment of local users who might not otherwise be represented;
9) to allow coordination of existing management entities;
10) to provide a salient example of how to achieve sustainable use of coastal/ marine resources.

---

## REFERENCES

1. International Union for the Conservation of Nature and Natural Resources (IUCN). Guidelines for Protected Area Management Categories. IUCN Gland, Switzerland, 1994.
2. Alexander, L.M. Large Marine Ecosystems: A new focus for marine resource management. Mar. Policy May 1993; 186-198.
3. Hatcher, B.G., R.E. Johannes, and A.I. Robertson. Review of research relevant to the conservation of shallow tropical marine ecosystems. Oceanogr. Mar. Biol. Ann. Rev. 1989; 27:337-414.
4. Towle, E.L. and C.S. Rogers. Case study on the Virgin Islands Biosphere Reserve. Case study, Workshop on Coastal Biosphere Reserves, 14-20 Aug 1989, San Francisco, CA. UNESCO.
5. van't Hof. Saba Marine Park: a proposal for integrated marine resource management in Saba. Netherlands Antilles National Parks Foundation (STINAPA), 1985.
6. Norse, E. Global Marine Biological Diversity. Washington, DC, Island Press, 1993.
7. Belleville, B. Diver-funded marine parks protect reefs and tourism. Rodales Scuba Diving, Nov./Dec. 1992:36-38.
8. Agardy, T. Guidelines for Coastal Biosphere Reserves. In: S. Humphrey, ed. Proc. of the Workshop on the Application of the Biosphere Reserve Concept to Coastal Areas, 14-20 August, 1989, San Francisco, CA, USA., IUCN Marine Conservation and Development Report 1992, Gland, Switzerland.
9. Hardin, G. Paramount positions in ecological economics. In: R. Costanza, ed. Environmental Economics, 1991: 47-57.

10. Kelleher, G. and R. Kenchington. Political and social dynamics for estab-
    lishing marine protected areas. Proc. of the UNESCO/IUCN Workshop
    on the Application of the Biosphere Reserve Concept to Coastal Areas,
    San Francisco, CA, 1989.
11. Kelleher, G.B. and R.A. Kenchington. Guidelines for Establishing Marine
    Protected Areas. IUCN Marine Conservation and Development Report,
    Gland, Switzerland, 1992.
12. Boo, E. Ecotourism: Potentials and Pitfalls. World Wildlife Fund, Wash.,
    DC, 1990.
13. Cousins, K. Ecotourism—Examples from the United States Coastal man-
    agement Programs and marine and estuarine parks. Coastal Zone 1991;
    '91: 1602-1610.
14. Kelleher, G. Identification of the Great Barrier Reef region as a particu-
    larly sensitive area. In: Proc. of the International Seminar of the Protec-
    tion of Sensitive Sea Areas, 1990: 170-179.
15. Kriwoken, L.K. The Great Barrier Reef Marine Park: an assessment of
    zoning methodology for Australian marine and estuarine protected areas.
    Maritime Studies 1987; 36:12-21.

# CHAPTER 6

# CLOSED AREA DESIGNATIONS AND HARVEST REFUGIA: ONE END OF THE TYPE SPECTRUM

## SECTION 1. CLOSED AREAS AND HARVEST REFUGIA

The use of closed areas in fisheries, also known as harvest refugia, presents an effective way to conserve stocks and habitats threatened by overexploitation, destructive fishing, and indirect degradation caused by pollution or the trickle-down effects of poor resource management in the vicinity. Such area closures may be established as a fisheries management tool, or may be a component of a wider array of spatially-defined management measures such as exist in a multiple use protected area, a biosphere reserve, or a coastal management plan. Among the benefits that the designation of closed areas accrue are conservation of stocks and species, maintenance of genetic diversity, protection of spawning stock biomass, reduction in growth overfishing, simplicity in being able to explain the management measure, relative ease of enforcement, provision of a baseline to monitor condition of stocks and the productivity or health of the ecosystem, and insurance against management failure.[1] By providing a means to target differential pressures applied to different stocks and different age groups, the closed area designation comes closest to approximating ecosystem-based, comprehensive management.[2] However, possibly high costs relating to exclusion of certain users, the logistical difficulties inherent in boundary delineation, scientific uncertainties relating to identification of ecologically critical areas, lost opportunity, and the spill-over of potentially increasing fishing pressure outside the limits of the closed area all necessitate that managers evaluate costs and benefits carefully before using closed areas to complement other forms of fisheries management.

---

It is likely that closed area designations of some form or another have been around since mankind first began harvesting the sea's bounty, making this one of the oldest forms of marine management. Area closures, whether formally designated or informally agreed to, have emerged as a way of settling user conflicts, conserving resources and laying claim to territory. Today closed area designations can be classified in at least four groups: 1) "traditional use" or "taboo" closed areas; 2) core areas within reserves or multiple use coastal plans; 3) harvest refugia for fisheries management; or 4) de facto area closures where exploitation is difficult or impossible due to physical constraints or poor resource availability.

Taboo closed areas evolve from patterns of historic use, usually in remote areas where cultural homogeneity exists, user conflicts are few, and traditional use is long-standing (e.g., see Johannes and McFarlane, 1991).[3] Cultural agreements to keep certain areas off-limits arise as an implicit form of self-regulation that prevents misuse of limited resources. In traditional societies, taboos are often directed at critical spawning events or ecologically sensitive areas.[4]

Closed areas that form one class of core areas within a multiple use protected area are also a means for targeting especially critical or sensitive areas. When core area designations are established for the intention of protecting productivity and biodiversity, they target important ecological processes for protection. Such processes can be physical, geochemical, or biological and include such things as upwelling, longshore and tidal fronts, warm and cold core rings, surface currents, freshwater mixing zones, nutrient loading, atmospheric exchange, population recruitment, keystone species, symbiotic associations, and predator/prey linkages.[5] The scientific basis for identifying such critical core areas exists—though harnessing that science to make it useful for management purposes requires synthesizing information across many disciplines and over a broad geographic scope.

Area closures that are designated specifically to protect "seed banks" or sources of recruits are becoming more and more common. Despite some semantical confusion in the literature on the definition of the term 'closed area' (ranging from areas which are deemed off limits for certain specified fisheries to areas closed to all users), for the purposes of this analysis I refer to such closures, when directed specifically at a fishery resource or the habitat that sustains it, as harvest refugia. Such refugia are being designated in the context of multiple use protected areas, coastal management plans, or as an independent fisheries management tool.

The primary benefit of using closed areas as a tool to complement other forms of resource management is the maintenance, or in some cases, boosting of fisheries productivity. Other corollary benefits include ease of management and reduced data collection needs, supplemental stocking, maintenance of 'control areas' for scientific research and monitoring, and potentially enhanced non-consumptive uses. It is also important to remember that until the last few decades there were many de facto refugia represented by areas beyond the technical or economic reach of resource users. These areas often are adequate to sustain the fisheries; unfortunately improved technology and an increasingly competitive economic climate has resulted in the destruction of these refugia.

Certain systems are better suited for the use of refugia. In general, the less dynamic the system, in terms of spatial and temporal variability, the more suitable. Coral reefs, for instance, are relatively static systems for which the precise location of certain features and resources at any given point in time is known.[6] Figure 6.1 shows a coral community underwater, emphasizing its relatively static nature. In contrast, highly dynamic ecosystems like those of temperate continental shelves have components that move about in often unpredictable ways.[7] For reasons relating to identification of critical areas, public education, and enforcement feasibility, use of harvest refugia is preferable in relatively "fixed" ecosystems.[8-14]

The growing use of closed areas or marine refugia has in part to do with relatively new information describing the explicit connection between certain coastal areas and maintenance of marine fisheries resources. As stated above, ecologically critical processes in nearshore ecosystems are often concentrated in areas that can be easily identified by physical parameters such as reef formations, extensive shallow water areas, certain types of coastal wetlands, continental shelf breaks and frontal systems. These are the types of areas that should serve as focal points for the establishment of systems or networks of small, discrete and easily identifiable closed areas.

## SECTION 2. NET BENEFITS OF CLOSED AREAS

Closed areas and harvest refugia are increasingly being selected from the portfolio of options available to marine resource managers, largely because conventional measures to manage fisheries and conserve marine ecosystems have repeatedly failed.[15] This failure entered the realm of public consciousness as the signs of mismanagement began to affect consumers as well as fishermen. Limiting fisheries management to controls on quantity of effort or catch ignores the potentially significant impact that fisheries activities have on ecosystems and their function. The use of spatio-temporal regulations, as made possible by area closures, ensures that the benefits of management are extended beyond just the target stock to wider segments of ecosystems themselves.[16] Thus closed areas, when used in conjunction with other forms of regulation, move fisheries management away from largely ineffective sectoral control to true conservation that benefits users as well as nature.[17,18]

The recent popularity of marine refugia has in part to do with scientific developments—reflecting where we currently stand in terms

Fig. 6.1. Coral reef community, Turks and Caicos Islands, West Indies. Photo by J. Spring.

of the science of population dynamics and identification of critical marine processes and areas.[16,19] The link between certain coastal areas and maintenance of marine fisheries resources has been clearly established (e.g., see Odum, 1984).[20] Ecologically critical processes in nearshore ecosystems are often concentrated in areas that can be easily identified by physical parameters such as reef formations, extensive shallow water areas, certain types of coastal wetlands, continental shelf breaks and frontal systems (for example see Haney, 1986; Loder and Greenberg, 1986; Steele, 1974).[21-23] The important biological processes that support fisheries productivity include spawning, migratory pathways, feeding, settlement and concentrated feeding.

Due to logistical considerations, economic efficiency, political infrastructure, and biases that western cultures seem to have difficulty shedding, the management of marine resources has historically centered on sectoral, top-down control. Such management practices ignore the linkages between various components in an ecosystem, their inherent vulnerabilities, the unpredictable nature of many ecosystem properties, and the impacts that resource use and exploitation have on non-target species. The result is a trend towards declining productivity and economic hardship, increased instability of ecosystems and rising discontent among users.

Fisheries regulation is increasingly brought up as an example of how science fails management. In the United States, a technically advanced country that devotes extensive resources to environmental protection and resource management, seventy percent of marine fisheries are said to be overexploited or exploited to maximum capacity, despite a decision-making structure that enables government to set quotas on effort and catch. One reason is that limiting fisheries management to controls on quantity of effort or catch ignores the potentially significant impact that fisheries activities have on ecosystems and their function. The use of spatio-temporal regulations made possible by area closures, however, ensures that the benefits of management are extended beyond just the target stock to wider segments of ecosystems themselves. Thus closed areas, when used in conjunction with other forms of regulation, move fisheries management away from ineffective sectoral approaches to true conservation that benefits all.

Fishing is therefore not necessarily a destructive practice and, in theory, management should be able to point all fisheries in the direction of sustainable and efficient use of these renewable resources. However, management assessments are rarely broad enough to include the comprehensive impact that the fishing activities have on the regenerative capability of the resource, leading to management decisions that are grounded in a myopic, and thus sometimes deceptive, view of what needs to be managed in order to maintain productivity and satisfy users.[24]

The extraction of fishery resources from the sea has direct and indirect consequences (see chapter 2). The impacts caused directly by the removal of the resource include diminished stock size and changes in resource distribution, possible overexploitation as the population size falls below the viable limit, changes in genetic composition that may affect evolutionary fitness, decreases in prey availability for those organisms that feed on the resource,[25] imbalances and restructuring of food web linkages,[26] and, in the extreme, biomass flips where the dominant components of the ecosystem are drastically reduced to minor members of the ecological community.[27]

Indirect impacts of fisheries activity are not a consequence of the quantity of resource removed but *how* it is removed. Such impacts include physical changes in habitat, lowered water quality as caused by chronic pollution or excessive nutrient loading, and episodic damage that result from boating mishaps, lost gear, etc.[24,28] Cumulative impacts, though difficult to assess, can occur when any of these effects are combined over time.[29]

## SECTION 3. CLOSED AREAS WITHIN MULTIPLE USE MANAGEMENT REGIMES

Focusing management on a single type of activity and a single stock at a time is largely ineffective, and increasingly more so as fishing pressures increase in scope and magnitude. The recognition and defense of public interest–including that of fishermen who represent the largest stakeholders—has often suffered as a result.[30] When closed areas are used to complement sectoral management, the chances of effectively maintaining the regenerative capacity of marine systems increases greatly. But, then, too—there are costs associated with excluding exploitative activity in an area. So where and when should harvest refugia be established?

There are many reasons why marine resource managers might choose to complement other management regimes with a system of fishery closed area designations. Harvest refugia can be formed to:

1) limit harvest of specific life stages (usually those critical to production or those especially vulnerable to direct and indirect effects of fishing activity);

2) prevent overexploitation of threatened stocks or species beyond replacement rates, or prevent growth overfishing;

3) protect sources of recruits or sinks for settlement (Agardy, 1995);

4) maintain the genetic composition and/or age structure of a stock or population (Polunin, 1983); or

5) buffer against management mistakes caused by scientific uncertainty or difficulties in executing management measures.[31,32]

The primary benefit of using closed areas as a tool to complement other forms of resource management is the maintenance, or in some cases, boosting of fisheries productivity. Other corollary benefits include ease of management and reduced data collection needs, supplemental stocking,[33] maintenance of 'control areas' for scientific research and monitoring, and potentially enhanced non-consumptive uses.[34]

From a fisheries management perspective, one of the most critical scientific considerations in the identification of refugia is where recruits come from and what affects their success.[35,36] Recruitment dynamics are often complex and seemingly unpredictable.[27,37] However, sources and sinks for recruits can be readily identified in some ecosystems (e.g., see Gaines and Bertness, 1992; James et al, 1990).[38,39] When scientific uncertainty is high or systems exhibit chaotic behavior, the use of multiple harvest refugia allows fisheries managers to hedge bets and increase the probability that productivity will be maintained.[40]

Where open access regimes for high value sedentary resources or highly territorial species leads to stock depletion, rotating harvest schemes can act as a modified closed area system.[41] If growth rates are slow and longevity high, only small proportions of the stock should be harvested annually. Use of a rotating harvest scheme rests on the assumption that stock range is divided into blocks or cells, each of which can be harvested in sequence. A further assumption is that adequate quota levels can be estimated; quota equalling the biomass of the resource divided by the number of cells. The period of harvest can be modified to allow one, or very few, year classes of optimal economic size to dominate the catch. This procedure has been widely used for other resources, in particular, forestry. As long as harvesting allows for residual spawners (or, more likely, that the stock in adjacent areas replenishes the population) the cell is given time to "rest" for a period of years or months. This may have both economic and ecological advantages.

Although the usefulness of closed areas and harvest refugia is being increasingly documented as resource managers turn to this management option, there are undeniable constraints to the broad applicability of this measure. Limited scientific knowledge on population replacement rates, dynamics, recruitment patterns, and impacts of fishing pressure on ecosystem function are all impediments to successful establishment of refugia.[42] The stochastic nature of many marine systems also undermines the usefulness of this approach, particularly if closed areas are treated as static and immutable entities instead of as flexible management measures.[43,44]

There may be social constraints limiting the applicability of closed areas as well. Fishermen are notoriously hard to regulate, precluding the acceptance of many new, potentially effective management tools. Closures having a scientific basis can be viewed by the fishing community as exclusionary practices that are somehow rooted in social discrimination–this predisposes user groups to reject the idea of area closures even before they have the chance to discover exactly why and how these would be beneficial to them.

Financial or logistical constraints that limit enforcement capability also limit the potential effectiveness of closed areas, although in many cases enforcement of designated closed areas may be more cost-effective than enforcement of quota limits, tow times, gear restrictions, etc. One important constraint is the physical demarcation of the closed area, so that its status is clear to users and prospective transgressors. The use of buoys can complement traditional latitude/longitude designations (Salm and Robinson, 1982), however, several researchers have raised the possibility that visibly demarcated areas may in fact become a lure to poachers.[45] Without giving enforcement feasibility due consideration, even the best-designed closed areas may be doomed to fail and further alienate fishing communities and other users.

If carefully planned and grounded in good scientific understanding of ecosystem dynamics, closed area designations can be an effective tool to complement other fisheries regulation. The prospect of increased management and enforcement that the implementation of such refugia entails will be a hard one to swallow for many members of the fishing community, but only until the effectiveness of such areas in maintaining and even increasing catch is demonstrated. Managers using this technique will have to be responsive to changes in scientific information, in the status of the resources, and in management needs in order to make refugia optimally effective. If they can do so, everyone stands to benefit from the use of this management measure.

## REFERENCES

1. National Marine Fisheries Service (U.S. Dept. of Commerce, National Oceanic and Atmospheric Administration). The potential of marine fishery reserves for reef fish management in the U.S. Southern Atlantic. NOAA Tech. Mem. 1990; NMFS-SEFC-261. SAFMC, Plan Development Team.
2. Agardy, T. Closed areas: a tool to complement other forms of fisheries management. In: K. Gimbel, ed. A Guidebook to Managing Fisheries Through Limited Access. Washington, DC, Center for Marine Conservation and World Wildlife Fund, 1993:197-204.
3. Johannes, R.E. and J.W. MacFarlane. Traditional Fishing in the Torres Strait Islands. CSIRO, Hobart, Tasmania, 1991.
4. Johannes, R.E., ed. Traditional Ecological Knowledge: A Collection of Essays. Gland, Switzerland, IUCN, 1989.
5. Agardy, T. Accommodating ecotourism in multiple use marine reserves. Ocean and Coastal Management 1993;20:219-239.
6. Salm, R.V. Ecological boundaries for coral reef preserves: principles and guidelines. Env. Cons. 1984; 11(3):209-215.

7. Agardy, T. and J.M. Broadus. Coastal and marine biosphere reserve nominations in the Acadian Boreal region: results of a cooperative effort between the U.S. and Canada. Proceedings of the Symposium on Biosphere Reserves, Fourth World Wilderness Congress, 14-17 Sept 1987, Estes Park, CO. U.S. Dept. of Interior, National Park Service, Atlanta, GA, USA. 1987.

8. Craik, W., R. Kenchington and G. Kelleher. Coral reef management. In: Dubinsky, ed. Ecosystems of the World. 1990 Vol. 25:453-466

9. Dahl, A.L. The challenge of conserving and managing coral reef ecosystems. UNEP Regional Seas Rep. 1985;69:85-87.

10. Kenchington, R.A. and E.T. Hudson. Coral Reef Management Handbook. Paris, France. UNESCO, 1984.

11. Livingston, R.J. Trophic organization of fishes in a coastal seagrass system. Mar. Ecol. Prog. Ser. 1982; 7:1-12.

12. McManus, J.W. The Spratly Islands: a marine park? Ambio 1994; 23(3):181-186.

13. Post, J.C. The economic feasibility and ecological sustainability of the Bonaire Marine Park, Dutch Antilles. In: M. Munasinghe and J.McNeely, eds. Protected Area Economics and Policy. World Bank and IUCN, Washington, DC, 1994:333-338.

14. Raines, P.S., J.M. Ridley and D. McCorry. The role of coral cay conservation in marine resource management in Belize. Presented at the World Parks Congress Workshops, 3-10 February 1992, Caracas, Venezuela.

15. Esping, L.E. The establishment of marine reserves. In: Proceedings of the International seminar on the Protection of Sensitive Sea Areas 1990: 378-386.

16. Davis, G.E. Designated harvest refugia: The next stage of marine fishery management in California. CalCOFI Rep. 1989; 30.

17. Ballantine, W.J. Marine reserves for New Zealand. Univ. of Auckland, Leigh Lab. Bull., 1991; 25.

18. Eldredge, M. The last wild place: marine reserves and reef fish. Washington, DC, Center for Marine Conservation, Washington, DC, 1994.

19. Davis, G.E. and J.W. Dodrill. Marine parks and sanctuaries for spiny lobster fisheries management. Proc. of the Gulf and Caribbean Fisheries Institute 1989; 32:197-207.

20. Odum, W.E. The relationship between protected coastal areas and marine fisheries genetic resources. In J. McNeely and K. Miller, eds. National Parks, Conservation and Development. Smithsonian Institution Press, Washington, DC, 1984.

21. Haney, J.C. Seabird affinities for Gulf Stream front eddies: Responses of mobile marine consumers to episodic upwelling. J. Mar. Res. 1986; 44:361-384.

22. Loder, J.W. and D.A. Greenberg. Predicted position of tidal fronts in the Gulf of Maine region. Cont. Shelf Res. 1986; 6(3):397-414.

23. Steele, J.H. The Structure of Marine Ecosystems. Cambridge, MA, Harvard University Press, 1974.

24. Dayton, P., R. Hofman, S. Thrush, and T. Agardy. Environmental effects of fishing. Aquatic Conservation: Marine and Freshwater Ecosystems 1995; 5:205-232.

25. Saila, S.B. and J.D. Parrish. Exploitation effects upon interspecific relationships in marine ecosystems. NOAA/NMFS Fish Bull. 1972; 70(2):383-393.

26. Goeden, G.B. Intensive fishing and a 'keystone' predator species: ingredients for community instability. Biol. Cons. 1982; 22: 273-281.

27. Fogarty, M.J., M.P. Sissenwine and E.B. Cohen. Recruitment variability and the dynamics of exploited marine populations. Trends in Ecology and Evolution 1991; 6(8):241-245.

28. International Council for the Exploration of the Sea (ICES). Report of the study group on ecosystem effects of fishing activity. Unpublished report of the Study Group of ICES, 1992.

29. Caddy, J.F. and G.D. Sharp. An Ecological Framework for Marine Fishery Investigations. FAO Fish. Tech. Pap. 1983; 283.

30. Agardy, T. Guidelines for Coastal Biosphere Reserves. In: S. Humphrey, ed. Proc. of the Workshop on the Application of the Biosphere Reserve Concept to Coastal Areas, 14-20 August, 1989, San Francisco, CA, USA, IUCN Marine Conservation and Development Report 1992, Gland, Switzerland.

31. Agardy, T. The Science of Conservation in the Coastal Zone: New Insights on How to Design, Implement and Monitor Marine Protected Areas. Proc. of the World Parks Congress 8-21 Feb. 1992, Caracas, Venezuela. IUCN, Gland, Switzerland. 1995.

32. Polunin, N. Marine genetic resources and the potential role of protected areas in conserving them. Env. Cons. 1983; 10(1):31-41.

33. Russ, G.R. and A.C. Alcala. Effects of intense fishing pressure on an assemblage of coral reef fishes. Mar. Ecol. Prog. Ser. 1989 56:13-27.

34. National Oceanic and Atmospheric Administration. Florida Keys National Marine Sanctuary: Draft Management Plan/ Environmental Impact Statement. Washington, DC, NOAA, 1995.

35. Csirke, J. Recruitment of the Peruvian anchovy (*Egraulis ringens*) and its and its dependence on the adult population. Rapp. P.V. Reun. CIEM 1980; 177:307-313.

36. Policansky, D. North Pacific halibut fishery management. In: National Research Council, Ecological Knowledge and Environmental Problem-Solving. Washington, DC, National Academy Press, 1986:138-149.

37. Holt, S.J. Recruitment in marine populations. Trends in Ecology and Evolution 1990; 5(7):231.

38. Gaines, S.D. and M.D. Bertness. Dispersal of juveniles and variable recruitment in sessile marine species. Nature 1992; 360:579-580.

39. James, M.K., I.J. Dight, and J.C. Day. Application of larval dispersal models to zoning of the Great Barrier Reef Marine Park. Proc. of PACON 90, 16-20 July 1990 Tokyo.

40. Carr, M.H. and D.C. Reed. Conceptual issues relevant to marine harvest refuges: examples from temperate reef fishes. Can J. Aquat. Sci. 1993; 50:2019-2028.

41. Caddy, J.F. Toward a comparative evaluation of human impacts on fishery ecosystems of enclosed and semi-enclosed seas. Rev. Fish. Sci. 1993; 1(1):57-95.

42. Soule, M.E. and D. Simberloff. What do genetics and ecology tell us about the design of nature reserves? Biol. Cons. 1986; 35:19-40.

43. Brodie, P.F., D.D. Sameoto and R.W. Sheldon. Population densities of euphasiids off Nova Scotia as indicated by net samples, whale stomach contents, and sonar. Limnology and Oceanography 1978;23:1264-1267.

44. Caddy, J. F. The protection of sensitive sea areas: a perspective on the conservation of critical marine habitats of importance to marine fisheries. In: Proc. of the International Seminar on the Protection of Sensitive Sea Areas, 1990:17-29.

45. Salm, R.V. and A.H. Robinson. Moorings for marine parks: design and placement. Parks 1982; 7(2):21-23.

# CHAPTER 7

# COASTAL BIOSPHERE RESERVES: THE OTHER END OF THE TYPE SPECTRUM

## SECTION 1. INTRODUCTION TO COASTAL BIOSPHERE RESERVES

The world's coastal zones are critical to humans (as well as the rest of the biosphere), yet these ecosystems continue to suffer acute degradation at the careless hands of man. The previous chapters have explained the form that this degradation takes, and have introduced the concept of multiple use marine protected areas as a way to combat threats to the oceans. Biosphere reserves, described in the following pages, constitute one class of multiple use protected area that represents the opposite end of the protected area type spectrum from small, narrowly targeted closed areas. The biosphere reserve concept is in fact an important tool for ensuring wise and sustainable use of natural resources in coastal areas.

Several researchers have described the functional units of the coastal zone. For example, Ray (1988) describes different types of catchment basin/receiving basin assemblages, together with their land and sea components.[1] These functional limits are determined by the bio-geophysical features of the system. Any conservation measure undertaken in one type of habitat should be developed with the entire functionally-delimited ecosystem in mind. The consideration of all landward and seaward components of the coastal ecosystem ensures that the target area is truly a functional unit, necessary for effective conservation irregardless of the type of habitat that is being managed.

Ecosystems in the coastal zone display a wide variety of characteristics, and provide diverse natural resources important to humans. Coastal habitat occurring in these ecosystems can be found on the margins of continents and around peninsulas, islands and archipelagos. Such coastal habitats include estuaries, fjords, sedimentary deltas, hypersaline lagoons, bays, marshes, mangroves, oyster reefs, worm and clam flats, submerged vegetation beds, channels, soft bottom communities, hard bottom habitats, rocky intertidal communities, cypress swamps, and coastal cliffs.

Offshore marine ecosystems are ecologically linked to the coastal zone and form an integral part of the coastal system itself. These offshore ecosystems may contain the following habitats: soft bottom benthic communities, kelp beds, coral and algal reefs, atolls, drowned reefs and submerged shorelines, submerged vegetation beds, sponge communities, shellfish beds, submarine canyons, sand banks and shoals, upwelling areas, deep holes, artificial reefs, continental slopes and shelves, seamounts, abyssal plains, ice environments and oceanographically-formed ephemeral phenomena such as fronts, gyres and currents.[2] These offshore habitat types, and the coastal habitats discussed above, are under various degrees of threat from human use, ranging from relatively pristine areas with minimal impact to fully urbanized or industrialized areas.

Given the broad definition of the coastal zone given above, it is recognized that coastal areas are unrivalled in importance in terms of biological diversity and productivity. Unfortunately, these areas are also the part of the biosphere where man's impact has been perhaps the greatest of all—a fact that makes coastal zone conservation as, if not more, exigent as on land. Conservation is made all the more urgent not only because more than two-thirds of the human population lives in the coastal belt but because the area boasts the world's highest rate of population growth.

The potential benefits that marine protected areas and biosphere reserves can confer are varied. All benefits—whether relating to the conservation of biologically important habitats, recovery and maintenance of overexploited stocks of resources, involvement of users in management decisions, or raising consciousness and education—can be traced back to two main features of multiple use areas: 1) the use of science to determine what must be protected when, and 2) the involvement of all sectors of society with vested interests in development of management and stewardship.[3]

Generally speaking, marine protected areas provide a way to focus conservation energies and management attention in order to ensure that existing or prospective human needs are met in an ecologically sustainable way. Biosphere reserves provide an advanced form of multiple use protected area design in which human beings are considered a vital component of the ecosystem. This "man-in" approach is manifested by bringing user groups directly into the planning and management process.[4]

In 1974, the United Nations' Educational, Scientific and Cultural Organization (UNESCO)'s Man and Biosphere (MAB) Programme launched an initiative to help develop special protected areas for inclusion in a global network of culturally and biologically important sites. To date over 500 officially sanctioned Biosphere Reserves exist in UNESCO's network; the many other protected areas that meet the program's criteria but that have not been officially nominated as biosphere reserves greatly expand this figure.

Coastal biosphere reserve design provides a useful model for integrating various forms of resource management in order to conserve marine systems while accommodating human needs. Although the biosphere reserve type of marine protected area has many of the same planning elements as more traditional protected area models (e.g., marine parks, reserves, sanctuaries) the crucial difference is the "man-in" approach to design and implementation.[5] A departure from simplistic, terrestrially-based models is clearly required due to peculiar problems that marine and coastal areas pose for conservation, such as: 1) the need to tackle the perception of global commons; 2) large scale biotic linkages that mean that impacts can reverberate through geographically vast areas; 3) conflicting jurisdictions; and 4) degrading forces that may occur far from the site of protection.

Coastal biosphere reserves are designed first and foremost to meet human expectations and needs, yet are ecological in their design (ensuring that critical ecosystem processes are protected). In addition to promoting better, more holistic conservation, biosphere reserves benefit humans by: 1) making the wider public aware of the value of the ecosystem and its resources; 2) providing an entry into a globally-linked system of reserves; 3) involving many segments of the local communities in planning, management and research/education; and 4) providing a salient example of how we should be managing coastal areas on regional and perhaps even global scales.

Biosphere reserves and their derivatives differ from traditional protected areas in that they tend to be larger than parks or sanctuaries, are zoned for many uses, include human communities within their boundaries, and stress education and research in addition to recreational values.[6] To meet UNESCO criteria (that is, to be included in the UNESCO network of officially sanctioned reserves), biosphere reserves must target three general roles: conservation, research and sustainable development.[7]

Biosphere reserves are planned under the auspices of Man and Biosphere (MAB) national committees, and nominations for each biosphere reserve must be approved by UNESCO for designation. Countries retain full sovereignty over their biosphere reserves, but MAB designation signifies a willingness to participate in the international network and to strive to meet the goals of the Biosphere Reserve Programme.[8]

Rather than acting in isolation as yet another (often superfluous) management agency, MAB national committees work with existing national, provincial and local agencies. In this way, biosphere reserves can act as an important aegis for regional planning and cooperation.

The UNESCO Biosphere Reserve Programme provides a useful example of how to incorporate human needs into long-term planning for conservation. Central to the model is multiple use zoning to protect sensitive habitats and critical ecological processes in core areas, while allowing managed use in buffer zones. This multiple use planning tool has particular potential in coastal areas, where conventional "garrison reserve" measures to preserve nature or protect the environment are not compatible with the open, multi-jurisdictional, and common property nature of marine systems.

The successful application of the biosphere reserve model in coastal areas will require a functional perspective that recognizes all the important linkages between and within marine and terrestrial areas. A functional approach allows delineation of the outer boundaries of the protected area (making the managed area a functionally viable entity), as well as helping to highlight where critical processes that drive the system are concentrated. If such "vital organs" of a system can be protected, humans will be able to continue to reap its resources and derive benefits from its use, leading to greater economic and sociological sustainability.

The early years of UNESCO's Biosphere Reserve Programme produced protected areas that showed little departure from traditional protected areas. Recognizing these shortcomings, UNESCO launched a strategic plan to make these reserves more effective. The 1984 Action Plan for Biosphere Reserves describes what the program sets out to achieve:

1) enhancing the role of the international network of reserves for ecosystem conservation;
2) improving and upgrading the management of existing reserves to correspond to their multi-purpose objectives;
3) promoting the conservation of key species and habitats;
4) promoting coordinated research projects in conservation science and ecology;
5) enhancing the role of reserves in regional planning;
6) promoting local participation in management of reserves;
7) supporting environmental education and training; and
8) using the network for information exchange on global biodiversity conservation and sustainable use.[9]

As the Biosphere Reserve Programme grows, it provides ever greater opportunities for fostering stewardship and promoting better, more holistic, resource management. Biosphere reserves have great potential for resolving user conflicts in the marine realm. The newly emerging coastal and marine reserves are also noteworthy in helping to achieve coordinated management with a strong footing in local communities.

Coastal biosphere reserves are in some cases the only feasible starting point for resolving user conflicts and establishing a basis for responsible use and attitudes. Reserves in this context are publicly recognizable spaces which allow users to become actively involved in planning rather than being recipients of management regimes imposed from outside. Users also partake in management–including undertaking enforcement of regulations–through partnerships between regulatory agencies and user groups.[10,11] Coastal biosphere reserves can provide the sociological basis for averting the 'tragedy of the commons' and fostering stewardship for ocean resource and ocean space among the people who most rely on healthy, intact coastal systems.

Shedding the old reputation of protected areas as elitist, unaffordable luxuries is difficult–especially in the eyes of user groups like fishing communities, for whom words like 'park' and 'reserve' have deeply ingrained negative connotations. Protected areas suffer from the fact that their benefits are hard to quantify and are often slow to be realized. We live in a world that craves instant gratification, and the mutually dependent functions of resource renewal, sustainability of ecosystem function, and long-term socioeconomic welfare of coastal peoples is not always linked in people's minds.[12] It is for this reason that clearly stated and specific management objectives for biosphere reserves and all other kinds of marine protected areas, against which progress can be quantitatively measured, are so critically important.

When is biosphere reserve establishment a more attractive alternative to traditional resource management, and where are coastal biosphere reserves preferable to marine parks or protected areas? There is no generic answer to this question, but there are clearly some circumstances when a coastal biosphere reserve designation makes more management sense than any other alternative. In countries without the administrative infrastructure necessary to mandate the creation of marine sanctuaries or parks, for instance, biosphere reserves may present a practical alternative. Similarly, states with long histories of resource conflict, left unresolved by conventional resource management, may turn to biosphere reserve planning for sustainable development. Other coastal nations or provinces with complex and productive ecosystems may find that biosphere reserves, with their large-scale, ecologically-based planning, are the only suitable means by which to address resource management on the scales appropriate to the system.

In practical terms, all the ways in which coastal biosphere reserves differ from other protected areas can be attributed to a single feature of coastal biosphere reserve planning: its flexibility.[13] Since the context for developing protected areas is vastly different in each place, the very character of each biosphere reserve will vary. It is because the biosphere reserve concept is widely applicable and can be adapted easily to fit specific needs that it is often a more suitable management option than the more rigid and conservative protected area approaches.

The following sections will provide detailed objectives and concerns addressed in biosphere reserve planning and attempt to show specifically how biosphere reserves become viable management alternatives.

## SECTION 2. OBJECTIVES AND ROLES OF COASTAL BIOSPHERE RESERVES

Coastal biosphere reserves differ from other protected areas in the nature of their management objectives. Coastal biosphere reserves must emphasize scientific monitoring, training, and information exchange, and fulfill their important role as a link in the global network of protected areas where standardized environmental information is collected and disseminated.[14] These reserves may also play an important conservation role in providing a voice to those users who find an entry into conventional resource management impossible. Since the success of coastal biosphere reserves, and indeed, the entire Biosphere Reserve Programme depends on the involvement and support of the local community, biosphere reserves provide a lasting forum for cooperative management of important marine resources and processes.

The development role of biosphere reserves is also a crucial concern in coastal areas. Given the high population densities in the coastal belt and the alarming rate of population growth in these areas, further and ever-increasing development of coastal resources and areas is inevitable. Long-term ecosystem health is often threatened by short-term economic gain, hence environmentally sound development must involve some compromises.[15,16] The way in which the development role is fulfilled in a particular coastal biosphere reserve will depend on the sociological character of the area bounded by the transition zone and cultural requirements of its inhabitants. Nevertheless, two generalizations about the development concern can be made: 1) in most developing countries (and some developed ones as well) the future of conservation is clouded unless local populations, who often have legitimate claims on the land and sea, recognize some benefit from proposed conservation measures; and 2) intelligent planning can lead to development that is both sustainable and economical in the long run.

Thus the development role of biosphere reserves can win supporters for the reserve and provide a wise management tool for coastal protection. Before these users become supporters of biosphere reserves, however, their rights (such as customary marine tenure) and needs must be recognized.[17]

The logistic role of biosphere reserves provided the first impetus for the establishment of the Biosphere Reserve Programme and remains an important, though perhaps secondary, concern.[11] The need for a worldwide network of natural areas in which standardized environmental monitoring and research are emphasized is now especially acute given threatening global change.[18] Coastal ecosystems provide the perfect venue for studying global changes in sea level, nutrient cycling, fate of greenhouse gases, primary production, etc. Thus coastal biosphere reserves can provide the scientific community with a network of existing sites for monitoring.[13] Other aspects of the logistic role of coastal biosphere reserves include facilitating information exchange and providing training opportunities. In order to meet these goals, it is important that fully functional biosphere reserves from all representative coastal environments be formed and included in the network.

It must be stressed that it is not sufficient for a coastal biosphere reserve to fulfill one of the above roles—fully all three must be evident for a biosphere reserve to be accepted into the MAB network and become a valid, viable biosphere reserve.[7] In many cases there are shortcomings in the implementation of the three-pronged approach which integrates conservation, development, and research. Often a site is a candidate for reserve status because of the necessity of conserving species in the area, but the activities and development needs of the local community are not brought into consideration.

The objectives of coastal biosphere reserves are necessarily diverse due to the differing circumstances and needs in each area where protected area planning takes place. Some of these objectives will be scientific, others conservation-oriented, and still others culturally-based. It is impossible to anticipate all of the objectives that planners might hope to achieve with a coastal biosphere reserve. Some objectives for coastal reserves include, inter alia:

1) providing a management focus for uses compatible with the Biosphere Reserve Programme without denying continued traditional use by local residents;

2) assuring the protection of species, habitats and processes critical to ecosystem maintenance and integrity;

3) assuring the protection of unique or endangered species, habitats, or cultures;

4) supporting the worldwide efforts of the Biosphere Reserve network and providing a representative ecosystem in the reserve network;

5) providing a coherent administrative mechanism for educational, scientific, and economic uses of coastal areas;

6) providing a mechanism for the resolution of conflicts arising from multiple uses of areas or resources;

7) reserving economically vital marine and coastal resources for the preferred use of residents or traditional users;

8) providing a voice for all interested parties, including those user groups not otherwise represented in the management process;

9) facilitating integrated research and education which focuses on the dynamics of coastal ecosystems and stresses the far-reaching connectivity between terrestrial and marine systems;

10) providing political incentives for the goal of sustainable use;

11)    demonstrating an illustrative model of integrated marine and coastal management on scales appropriate to the ecology of the system; and

12)    providing an example to administrators, politicians, lending agencies, etc., of how integrated management which takes into account the needs and knowledge of local people can guarantee the survival of fragile coastal ecosystems and the cultures that depend on them.

## SECTION 3. ZONING WITHIN COASTAL BIOSPHERE RESERVES

### 3.1. GENERAL ZONING PRINCIPLES

The underlying basis for conservation within a biosphere reserve is ecosystematic–such that areas identified as critical, sensitive, or important to threatened species are strictly protected in core zones.[6] Around these protected core areas are buffer zones, in which human activities which might adversely impact the core are controlled. Beyond these core and buffer limits, biosphere reserves are bounded by vast areas known as transition zones or areas of cooperation. All biosphere reserves in the MAB Programme adhere to these general design criteria, as elaborated by UNESCO in 1974.[8]

Although focusing on the land-sea interface, coastal biosphere reserves can also include seemingly unconnected areas on land (for example, watersheds) and in the open sea (for example, upwelling areas). Coastal biosphere reserves also must be comprised of different zones for different uses or levels of use, including core areas on land and in the sea and regulated use (buffer) areas to allow for long-term development.

Biosphere reserves should be designed to ensure that populations, habitats, and ecological processes, as well as threatened species, are conserved. This will require that core areas target not only the obvious (and most commonly targeted) components in the ecosystem, but the more complex and less obvious, as well. Buffer areas have to reflect this ecologically-based conservation and ensure that anthropogenic impacts are minimized.

As stated numerous times in this volume, ocean and coastal systems are large-scale, dynamic and poorly understood. Coastal areas share characteristics of both the terrestrial and marine systems they interface with, and are often characterized by higher species diversity and productivity than neighboring systems.[1] Due to this intrinsic complexity, the management of coastal areas may require more extensive assessment and planning than terrestrial park management.

The characteristics of coastal areas that have important implications for management vary from region to region. In tropical areas with abundant coral reefs, for instance, most of the critical resources can be delimited on a map, having relation to the substrate (Hatcher et al, 1989).[19] Thus, if a tropical fishery is threatened by overharvesting, management can be easily legislated by restricting fishing in a given area(s). Sometimes the regulations call for closing an area to fishing or other resource extraction entirely–creating a harvest refugium within the bounds of the reserve (see chapter 6). This regulated area then becomes closely analogous to resource protection on land. If, however, such management is needed in temperate or deep-water areas, where resources vary widely in time and space and have little linkage to the substrate, delimiting a management area becomes more difficult.[12] In effect there is little connection between the water column with its dynamic components and the benthos.[11] Furthermore, in coastal areas where oceanographic phenomena such as warm-core rings abruptly and unpredictably change the structure of the ecological community, management becomes even more complex.

What this suggests is that the design criteria for coastal reserves will be different in different geographic regions (reiterating the value of a flexible design). But it does not mean that zoning in areas of high dynamism is untenable or undesirable, or that biosphere reserve planning should be abandoned in favor of traditional management.[20] Zoning plans may well require more intensive investigation into both the dynamics of the ecosystem and the patterns of human use within it in such areas, but the end result will be a method of management which satisfies as many users as possible while conserving the ecosystem.

However designed, coastal biosphere reserves must be comprised of core, buffer, and transition zones to satisfy the criteria for their inclusion in the reserve network of UNESCO's Man and Biosphere Programme.

## 3.2. CORE AREAS

Core areas are the focal points of the biosphere reserve, highlighting ecologically important habitats or processes that are potentially threatened by unmitigated development. The core area functions primarily for conservation and its ecological integrity must not be compromised. This does not exclude the possibility that different types and levels of use that are seen to be compatible with the conservation objectives of the reserve will be permitted within the core area.

The nature of the core areas will reflect the nature of the conservation and other objectives of the particular biosphere reserve where it is found. These objectives must be clearly stated before biosphere reserve planning commences. The number of core areas will be determined by what is needed to meet the overall conservation, development, and research objectives of the reserve. Because of the patchy distribution of species and habitats in coastal areas and the linkages between them, it is likely that multiple cores will be needed in most coastal biosphere reserves. The natural linkages among coastal, marine and inland realms preclude the effective management of any of these areas independent of their adjacent or linked habitats.[10]

Core areas in coastal biosphere reserves should encompass terrestrial and marine components. This may be difficult to achieve given the nature of coastal management in general. According to Batisse (1989), most terrestrial ecologists tend to forget the marine part of the coastal zone and oceanographers seem to consider that it stops at the high tide mark.[10] This forceful split in scientific disciplines and institutions is aggravated in practically all countries of the world by an equal split between responsible administrations. Thus the identification and designation of marine and terrestrial core areas will likely be difficult, but key to the efficacy of coastal biosphere reserves.

There are important logistic objectives that the core areas will also help fulfill. An international network of coastal biosphere reserves will provide a series of protected, representative habitats in which comparative research and monitoring can be undertaken. Thus, in designating core areas, planners should consider the range of research programs which might be initiated once the reserve has been designated.

The boundaries of core areas will reflect the degree of dynamism in the coastal area. Some cores will be spatially discrete with fixed boundaries, such as small island cores or cores in coastal lagoons. In other cases, for instance where there are seasonal concentrations of species or processes, it may be possible to have a dynamic core.[20] An example of a dynamic core might be an offshore area which provides seasonal habitat for spawning fishes. The conservation objectives of the reserve may be fulfilled in part by the administration of strict protection during this spawning season—hence the core would only be a seasonal entity. Such dynamic cores may be defined by species concentrations, hydrological or oceanographic patterns, ecotone boundaries, levels of resource use and administrative considerations. In general, core boundaries must encompass enough habitat to enable the system to be self-sustaining and to sufficiently meet the conservation goals of the reserve.[6]

The question of core area requirements in coastal biosphere reserves has sparked some controversy in recent discussions.[11] The focus of these discussions has been to identify the minimum degree of protection which must be directed at an area before it can meet the requirements of a core area. In terrestrial biosphere reserves, core areas must be strictly protected by law: either by existing national legislation or by special proclamation and/or mandate (as in the case of Madagascar and Mexico where Biosphere Reserves have been incorporated into national law). Since coastal biosphere reserves can and should include marine core areas, which are never proffered the same degree of protection as terrestrial cores, the requirements for marine core areas are ambiguous. In most cases, the existence of an Exclusive Economic Zone (EEZ) or territorial sea designation is sufficient to indicate that the nation has a commitment to protecting important core areas. In areas lacking EEZ designations, legislative mandate may be required to ensure core areas will be protected to the fullest extent possible.

### 3.3. BUFFER ZONES

Core areas must be fringed by buffer zones which secure the integrity of the core and protect its ecological linkages. Buffer zones are areas of regulated use in which the aim of management is to conserve the core. Critical ecological processes which impact and regulate the core systems must be identified before the limits and regulatory nature of the buffer areas can be determined.[6,13] Those processes which affect the maintenance of the core system must be conserved in the buffer zone. The buffer thus functions as exactly that: it buffers the core from external negative impacts.

Buffer zones also have value in demonstrating the multi-purpose goals of the biosphere reserve, including the development and application of an institutional infrastructure that resolves agency conflicts and integrates management activities among the various agencies with jurisdiction in the coastal realm.

The boundaries of buffer zones will be primarily determined by socio-political considerations. In other words, the shape and size of any area of regulated use will reflect the managerial capacity of the agencies and organizations which govern the use of the resource. The boundary is also influenced by pre-existing or potential patterns of human use. Thus, in setting out to demarcate buffer zone boundaries, planners will have to assess the patterns of human use in the area and the extent to which those patterns can be controlled and/or monitored.

### 3.4. TRANSITION AREAS

The transition area, that vast area that proscribes the operational limits of the biosphere reserve itself, is a more or less nebulously defined area of cooperation. It is the intent to cooperate and jointly manage resources that provides the impetus for establishing the transition area. This cooperation should bring together existing governmental agencies, non-governmental organizations and interested citizens as a management team.[5]

The transition area is a place to demonstrate the use of regulatory mechanisms to control activities, safeguard environments, and manage species' exploitation where these affect the goals of the biosphere reserve. Thus the transition area is defined in space by regulatory mechanisms and can expand as other interested parties join the biosphere reserve for cooperative activities.[8]

To bring the coastal biosphere reserve into the realm of national development and planning, regulatory mechanisms are required that link the reserve with the existing planning process. A coastal biosphere

reserve that is superimposed on other planning and development initiatives, with little forethought as to where and how the reserve will
fit into the existing framework, has a low probability of long-term
success. Therefore it is crucial to consider the issues encompassed below before drawing up the final design and management plan of a
biosphere reserve:

1)   Are important ecological processes and life-sustaining sys-
     tems included in core zones to the fullest extent possible?
2)   Are important natural areas included within the transition
     zone and limits of the reserve?
3)   Are areas with potential for both education and research
     included in the reserve?
4)   Are areas of social, historic and economic importance in-
     cluded, and are local needs emphasized?
5)   Are demonstrable examples of "harmonious landscapes" or
     sustainable use patterns present?
6)   Have the major biogeophysical components of the targeted
     ecosystem been identified, and do zoning boundaries re-
     flect these ecosystem realities?
7)   Are ecosystem processes well understood and have linkages
     been identified?
8)   Have existing resource use patterns been assessed, and all
     users identified?
9)   Have economic evaluations of important resources been un-
     dertaken, and if so, have they been used to justify zoning
     for long-term sustainable use?
10)  How will the operational success of the zoning scheme and
     the coastal biosphere reserve itself be monitored?

One might conclude from the preceding discussion that complete
scientific knowledge is a necessary prerequisite for biosphere reserve
zoning and design. As stated in chapter 2, this is not the case. Some
level of knowledge about the ecosystem and its critical processes is a
requirement, otherwise biosphere reserve planning cannot be said to
be ecologically-based. But complete scientific knowledge is never available
to planners, and the urgent necessity of managing coastal areas now
requires us to work with what knowledge we have. Again, it is of ut-
most importance that coastal biosphere reserves, and the zoning within
them, remain flexible. As new information emerges and scientific knowledge becomes more fine-tuned, adjustments can be made to alter the
number, shape and size of core areas or the regulations within the
buffer zones. And as more information is communicated over greater
and greater distances, the zone of cooperation or transition area can
be extended to incorporate a wider cooperatively-managed biosphere
reserve area.

## SECTION 4. A GENERAL PROTOCOL FOR PLANNING COASTAL BIOSPHERE RESERVES

### 4.1. ELEMENTS OF THE PLANNING PROCESS

Planning effective coastal biosphere reserves requires sufficient knowledge about the biogeography of the region and the ecological processes which are critical to its integrity and character, as well as an understanding of existing patterns of resource use and the needs of local peoples. There is certainly no correct way of planning and implementing coastal biosphere reserves, but there are a number of key points which must be considered during planning.

The planning process is at the heart of the successful translation of MAB philosophy and the biosphere reserve concept into meaningful management. In general, coastal planning is the deliberate action between knowledge (in other words, the scientific and objective understanding of environmental problems) and implementation (in other words, the effective resolution of those problems).[5] How this is achieved will vary with socio-political settings and with the nature of the environmental problems which confront managers.

Coastal biosphere reserve planning will always require five basic steps: 1) the recognition of need; 2) resource and resource use assessments; 3) an analysis of the feasibility of establishing and maintaining a viable coastal biosphere reserve; 4) nomination of the reserve for inclusion in the international biosphere reserve network of the MAB Programme; and 5) implementation of the plan.

### 4.2. RECOGNITION OF NEED

What initiates the planning process and acts as a catalyst throughout is the recognition of threats and the necessity of mitigating against them. Without a perceived problem, there is little impetus to control use or develop a protected area. The identification of need for a coastal biosphere reserve, in particular, can come from a variety of sources: from resource users, special interest groups, educational institutions, government agencies, or MAB National Committees.

MAB National Committees should be proactive in identifying opportunities for establishing coastal biosphere reserves, but they should also encourage identification of potential coastal biosphere reserve sites by other groups as well (UNESCO, 1984).[9] Some obvious candidates for coastal biosphere reserve planning are existing biosphere reserves situated near coastlines, which can be geographically extended to include entire coastal systems. Other potential sites are existing protected areas where a wider cooperative basis for management is needed, or where coordination between the management of various components of the system is required.

The first step in the coastal biosphere reserve planning is thus recognizing the need for special area management. Once the need for coastal management is recognized, some mechanism is required by which all the actors concerned with environmental protection, local and regional development, or scientific research and monitoring be brought together to work towards cooperative agreements for planning and implementing the future biosphere reserve. Here, it is understood that the participation of each institution or interest group in the planning and implementation of the future biosphere reserve will be governed by the legal mandate or mission of each institution or group. In other words, there is no "take over" of one institution of another: all participating bodies will seek to incorporate the objectives of the future biosphere reserve within their own planning processes and management regimes to the maximum extent possible.[21]

The mechanism whereby this cooperation can be obtained will of course differ from region to region around the world. In some cases, the preparation and signing of a cooperative agreement will suffice. In other cases, it may be more opportune to designate a selection panel to oversee the coastal biosphere reserve planning process. Such a panel should be equitably composed of scientific experts, resource managers, representatives from educational institutions and resource users.

Any Biosphere Reserve Panel will likely have preconceived notions about the goals that a coastal biosphere reserve should accomplish, since, in most cases, efforts to designate a coastal reserve will arise in response to a stated need. Nonetheless, the first task of a panel will be to discuss those needs and create a detailed list of objectives for the project.

The importance of identifying specific objectives for the biosphere reserve cannot be understated. The nature of the biosphere reserve itself will be a reflection of this statement of need; if it is not, the coastal biosphere reserve will fail to fulfill its roles.

## 4.3. RESOURCE AND RESOURCE USE ASSESSMENT AND FEASIBILITY STUDY

Once a panel has identified the specific objectives for biosphere reserve planning, three tasks should be undertaken simultaneously: 1) a biogeographic review of the region and its resources; 2) a systematic assessment of the uses of those resources, and their economic and social value; and 3) an analysis of the type and effectiveness of existing legislation aimed at protecting those resources. These three assessments will form the basis for a feasibility study, which is the backbone of the coastal biosphere reserve nomination report and the planning process.

The feasibility study must be comprehensive to be realistic. As such, it must include an overview of the region, including its physical features, ecosystem dynamics, and sociological character. The study should present a detailed assessment of resources in the areas targeted as possible core and buffer zones, and should explain, to the fullest extent possible, the factors which contribute to the maintenance of the ecosystem and the regeneration of key renewable resources. Some of this assessment can be accomplished with mapping, and if possible, the use of computerized databases and geographic information systems (GIS).[22] Also included should be a list of resource use issues of regional and local significance, including land use conflicts, ocean space use and access, socio-economics and environmental concerns. In many cases, issue identification will be best achieved through public hearings.

The feasibility study is more than a supporting document for the panel's final nomination report.[21] It should be an objective assessment of the prospects for the reserve's implementation over the long-term. Such an assessment must determine the nature and levels of interest in developing a reserve plan and in practicing cooperative management under the aegis of MAB. If little willingness to support a reserve effort is discovered, or if the political prospects are such that the reserve would have little chance of accomplishing its aims, then the panel should look at alternative solutions for coastal management.

## 4.4. BIOSPHERE RESERVE NOMINATIONS

When assessments, feasibility studies, and nomination reports have been prepared, the panel or identified lead agency in the case of a less structured cooperative agreement, should submit a formal report to its MAB National Committee. In the event that there is no MAB National Committee, such reports should be submitted to the designated focal point for the MAB Programme or the UNESCO National Commission (the national body responsible for cooperation with UNESCO). In turn, the MAB National Committee submit the nomination dossiers to the MAB Secretariat at UNESCO Headquarters in Paris (France), which presents biosphere reserve nominations to the Bureau of the MAB International Coordinating Council for approval. This Bureau meets usually once per year. The sites approved as biosphere reserves under the MAB Programme receive a formal biosphere reserve certificate signed by the Director-General of UNESCO.[8]

If the coastal biosphere reserve nomination is approved and the biosphere reserve is designated, a plan for implementation must be created. This plan may or may not be a formal document, or may a part of the feasibility study, depending on national requirements. In any case, it is imperative that a strategy to follow through on the planning process to put theory into action be considered.

## SECTION 5. IMPLEMENTING AND MANAGING BIOSPHERE RESERVES

A coastal biosphere reserve should promote collaboration and interaction of local people and agencies with responsibilities in the area, and must attempt to facilitate the realization of local aspirations and needs. Due to the linkages between marine and terrestrial components of coastal biosphere reserves, specific coordinating mechanisms must be established. However, these coordinating mechanisms should not be superimposed on existing management entities without their involvement in the planning process. The creation of new institutions, legislation, and regulations should be minimized.[23] Instead, existing instruments for management should be used to the maximum extent possible, except of course where overwhelming public support for new legislation exists. In some cases supplemental enforcement provisions may be necessary; this is especially true where new legislation to protect resources is mandated.

Communication among all those involved in the coastal biosphere reserve is an essential tool for implementation. Management should be supported through educational programs to ensure that those affected are aware of their rights and responsibilities under the reserve.[5] This will act to encourage community support for the objectives of the coastal biosphere reserve even further, since a well-designed education and public involvement programme can generate political and societal enthusiasm. This, in turn, generates pride and commitment, two elements which experience has shown to be critical in the long-term success of any conservation endeavor. Figure 7.1 illustrates that community involvement in park planning should include even those sectors of society not traditionally involved in decision-making.

In order to ensure that the coastal biosphere reserve is truly meeting its objectives over the long-term, regular monitoring of compliance and public support should be undertaken. This monitoring should determine the extent to which management plans are being followed, the condition of managed ecosystems and important resources within the reserve, and the socio-political climate for management. Periodic evaluations of management plans and implementation will allow resource management to become adaptive and responsive to changing conditions. As new scientific information is accumulated and as resource use patterns change, the zoning of the reserve or the objectives of management may have to change as well.[6] Monitoring the implementation of management plans over long time horizons is the only way that coastal biosphere reserve planning can lead to sustainable use of the environment.

More so than perhaps all other forms of protected area management, the implementation and long-term administration of coastal biosphere reserves varies according to circumstance. Since coastal biosphere reserves must be formed in the administrative context of existing national legislation, local governances, and international treaties, the form of management which the reserve adopts will always be a function of the specific conditions to which it is tailored.

Fig. 7.1. Members of the community involved in the drafting of a marine park plan, Tanzania. Photo by T. Agardy.

It is thus difficult to envisage a generic model for management of coastal biosphere reserves. In the United States, for instance, coastal biosphere reserve management must fit in a legislative framework comprised of existing national laws such as NEPA (the National Environmental Protection Act), CZMA (Coastal Zone Management Act) and the Clean Air and the Clean Water Acts, among others. Depending on the scope of the coastal reserve, these and other laws are the jurisdiction of city councils, state environment and wildlife agencies, the National Oceanic and Atmospheric Administration, the Environmental Protection Agency, Fish and Wildlife Service and other national bureaus and agencies. Depending on the specific circumstances, the long-term management of a coastal biosphere reserve in the United States will likely be under the governance of an overseeing council composed of representatives from these agencies and councils, coordinated by one or more full-time administrators of the reserve. This is, in fact, the case in the few existing coastal biosphere reserves in the U.S., including the Californian Coastal Biosphere Reserve, for instance.

In other countries with a less rigid framework for coastal resource management, the implementation and long-term administration of coastal biosphere reserves may be more centralized. The nature of the overseeing body may be para-governmental or non-governmental, or may exist as a newly formed office within an existing government agency. Ultimately, the nature of the reserve's administration will reflect why that particular coastal biosphere reserve was established and which sectors of society are its most active proponents.

The necessity of maintaining flexibility in the planning of coastal biosphere reserves has already been stressed. This flexibility must be sustained, beyond implementation of the reserve and the establishment of a governing council or other overseeing body. As new information about resource capabilities and ecosystem function is gained, and as local people's needs change, the administration of the reserve and the specific rules which proscribe their use will have to be tailored.

## SECTION 6. RESEARCH ORIENTATIONS WITHIN COASTAL BIOSPHERE RESERVES

Research and monitoring are integral components of functional coastal biosphere reserves. The four main research orientations for the MAB Programme are also applicable to coastal reserves. Indeed, the orientations attempt to reflect the fact that human issues and social sciences are just as important as ecological considerations when trying to solve natural resource use problems. These research orientations are briefly described below, highlighting aspects for study in coastal biosphere reserves.[8]

## 1) ECOSYSTEM FUNCTIONING UNDER DIFFERENT INTENSITIES OF HUMAN IMPACT

This is the more traditional approach to ecological studies but is of vital importance in coastal systems where, as was mentioned earlier, there is a relative lack of knowledge of how coastal systems function in the natural state and what impacts human activities have on ecosystem structure and function. The understanding of land-sea coupling processes is clearly important to better manage the system as a whole. Indeed, while terrestrial ecosystems on land are thought to be influenced strongly by biological interactions, in contrast, marine systems are believed to be governed more by physical and chemical processes. The coastal zone, comprising both land and sea, is clearly intermediate. Studies in coastal biosphere reserves should also focus on the effects of terrestrial activities on the nearshore marine environment.

## 2) MANAGEMENT AND RESTORATION OF HUMAN-IMPACTED RESOURCES

Coastal ecosystems are undergoing increasing degradation in many parts of the world and it is therefore important to study the limits of coastal ecosystem resilience. Of particular concern are: i) how much disturbance (including loss of biological diversity) can coastal ecosystems withstand before their structure and function are impaired; and ii) once degraded, how quickly (if at all) can a coastal ecosystem recover to a "normal" state. Research into the restoration, for example, of mangroves through transplantations has been carried out in several parts of the world and should be further encouraged. The creation of artificial environments using man-made structures (for example artificial reefs based on concrete pyramids or old tires) can be of great value since such structures attract fish and may be caught and improve local revenue. Such artificial structures may not always succeed and may even have negative effects such as through increased sedimentation, hence the need to better understand the ecological processes involved in creating artificial environments.

## 3) HUMAN INVESTMENTS AND RESOURCE USE

This orientation considers the linkages between economics, human welfare and ecological sustainability as a result of local, regional and global forces. One particular aspect of interest in coastal biosphere reserves is the estimation of the direct economic value of coastal resources, for example, from mangroves, fisheries, and clean beaches for tourism, which are a graphic means of demonstrating the value of natural resources to planners and development agencies. The indirect value of coastal resources should also be considered. Coral reefs, for example, have been aptly termed "self repairing breakwaters" as a result of their capacity to protect shorelines. The economic value of protecting natural resources (as opposed to direct exploitation, e.g., of a coral reef for building material) should be included in economic analyses of development projects.

## 4) Human Response to Environmental Stress

In coastal biosphere reserves, studies address societal responses to adverse influences on coastal systems, for example, from heavy metal accumulation (such as in the case of mercury and the minamata disease), fertilizers, sewage (municipal, or detritus from intensive fish farming), noise (airports), etc. Social science research could also be linked to this orientation, for example, comparing the different legal, institutional and management regimes in coastal areas. The socio-political and legal issues arising at the land-sea interface are unique and little understood and need to be addressed by appropriate research.

It is clear that the particular research emphasis of a given coastal biosphere reserve will vary from region to region depending on local requirements. The MAB approach of integrated systems research, combining the natural and the social sciences, basic and applied studies, with comparisons with other sites, is particularly appropriate to coastal systems.[8] This approach will help to better understand the functioning of a coastal area as a whole system, to monitor its evolution over time, predict changes and propose remedial measures. Such work is by definition complex and time consuming but is well worthwhile, not only for better protecting and managing the coastal biosphere reserve under study, but also for providing information which may be used to solve conservation and management problems in other sites in other parts of the world. Finally, it should be emphasized that a successful coastal biosphere reserve will serve a very important demonstration role to illustrate how conservation, scientific research and monitoring and rural development can synergistically combine for the benefit of present and future generations.

The international network of biosphere reserves includes a diverse assemblage of habitat types and human communities. In each, three major goals of the Biosphere Reserve Programme are being met: 1) conserving the diversity and integrity of biotic communities for present and future use and safeguarding the genetic diversity of species; 2) providing areas for ecological and environmental research, including baseline monitoring; and 3) providing facilities for education and training. This emphasis on combining multiple functions within a single site differentiates biosphere reserves from parks or other protected areas and ensures long-term public support of both the concept and individual reserves. More and more, countries are turning to biosphere reserve planning as a way to ensure the conservation of their important natural resources while at the same time encouraging continued sustainable human use.

Biosphere reserves present an important new resource management option to governments and planners. Their departure from traditional forms of management means that biosphere reserves can be more flexible and adaptable to local needs and conditions, and that biosphere reserve planning can fill in voids in conservation and resource management heretofore left unfilled. One such void is in the comprehensive and integrated management of coastal ecosystems.

Several general points may be made in summarizing coastal biosphere reserves. First and foremost, their design must reflect the ecological realities of linkages between habitats on both the landward and seaward sides of the coastline. It is ineffective to have a coastal biosphere reserve which considers only a few of the components of the coastal system. Second, coastal biosphere reserves will only be successful if the reasons for establishing the reserve are clearly spelled out, and if all interested parties are an intrinsic part of the planning process. These objectives, be they oriented towards conservation, sustainable development, or research and monitoring, will define the nature of the reserve and influence its design. Third, coastal biosphere reserves must utilize the principle of zoning for multiple use, such that ecologically critical areas are left undisturbed as core zones and use is controlled in buffers. Fourth, the vital role of coastal biosphere reserves in coordinating management and providing a much-needed forum for cooperation should not be overlooked. Finally, coastal biosphere reserves must remain flexible and adaptive, changing as conditions change. They are wheels to be set in motion, dynamic yet staying on track to meet their many objectives.

## REFERENCES

1. Ray, G.C. Sustainable use of the ocean. In: Changing the Global Environment. New York, Academic Press, 1988:71-87.
2. Linden, O. Oceanographic features of importance for coastal marine biosphere reserves. SFCBR, 1989.
3. Agardy, T. Accommodating ecotourism in multiple use marine reserves. Ocean and Coastal Management 1993; 20:219-239.
4. Fiske, S.J. Sociocultural aspects of establishing marine protected areas. Ocean and Coastal Management 1992; 18:25-46.
5. Salm, R.V. and J.A. Dobbin. Management and administration of marine protected areas. In: Proc. of the Workshop in the Application of the Biosphere Reserve Concept to Coastal Areas, 14-20 August 1989, San Francisco, CA, USA.
6. Agardy, T. Guidelines for Coastal Biosphere Reserves. In: S. Humphrey, ed. Proc. of the Workshop on the Application of the Biosphere Reserve Concept to Coastal Areas, 14-20 August, 1989, San Francisco, CA, USA, IUCN Marine Conservation and Development Report 1992, Gland, Switzerland.
7. Batisse, M. Development and implementation of the biosphere reserve concept and its applicability to coastal regions. Envir. Cons. 1990; 17(2):111-116.

8. Robertson-Vernhes, J. Biosphere Reserves: the beginnings, the present, and future challenges. In: W. Gregg, ed. Proc of the Symp. on Biosphere Reserves. Fourth World Wilderness Congress, 14-17 Sept. 1987, Estes Park, Colo. USA, 1989.

9. United Nations Educational, Scientific and Cultural Organization (UNESCO). Action Plan for Biosphere Reserves. UNESCO, Paris, France, 1984.

10. Batisse, M. Development and implementation of the biosphere reserve concept in coastal areas. Proc. of the Workshop on Coastal Biosphere Reserves, 14-20 August 1989, San Francisco, CA. Paris, UNESCO, 1989.

11. Kenchington, R.A. and M.T. Agardy. Achieving marine conservation through biosphere reserve planning. Env. Cons. 1990; 17(1):39-44.

12. Pernetta, J. and D. Elder. Cross-sectoral, Integrated Coastal Area Planning: Guidelines and Principles for Coastal Area Development. IUCN Marine Conservation and Development Report, Gland Switzerland, 1993.

13. Agardy, T. The Science of Conservation in the Coastal Zone: New Insights on How to Design, Implement and Monitor Marine Protected Areas. Proc. of the World Parks Congress 8-21 Feb. 1992, Caracas, Venezuela. IUCN, Gland, Switzerland. 1995.

14. Clark, J.R. Management of coastal barrier biosphere reserves. Bioscience 1991; 41(5):331-336.

15. Brundtland, G. Our Common Future. In: V. Martin, ed. For the Conservation of the Earth. Golden, Colo. Fulcrum Pres, 1988:8-12.

16. Clark, C. Clear-cut economies. The Sciences, NY Academy of Science, Winter 1988.

17. Johannes, R.E. Marine conservation in relation to traditional lifestyles of tropical artisanal fishermen. The Environmentalist 1984;4(7).

18. Markham, A. Potential impacts of climate change on ecosystems: a review of implications for policymakers and conservation biologists. Cli. Change Res. 1995 6:179-191.

19. Hatcher, B.G., R.E. Johannes, and A.I. Robertson. Review of research relevant to the conservation of shallow tropical marine ecosystems. Oceanogr. Mar. Biol. Ann. Rev. 1989; 27:337-414.

20. Agardy, T. and J.M. Broadus. Coastal and marine biosphere reserve nominations in the Acadian Boreal region: results of a cooperative effort between the U.S. and Canada. Proceedings of the Symposium on Biosphere Reserves, Fourth World Wilderness Congress, 14-17 Sept 1987, Estes Park, CO. U.S. Dept. of Interior, National Park Service, Atlanta, GA, USA. 1987.

21. Gregg, W. Draft guidelines for biosphere reserves and regional MAB programs in the U.S. Unpublished document, U.S. Park Service, Wash., DC, 1989.

22. van Claasen, D., ed. The application of digital remote sensing techniques in coral reef, oceanographic and estuarine studies. UNESCO, Paris, 1985.

23. Kelleher,G.B. and R.A. Kenchington. Guidelines for Establishing Marine Protected Areas. IUCN Marine Conservation and Development Report, Gland, Switzerland, 1992.

# CHAPTER 8

# MARINE PROTECTED AREA CASE STUDIES

## SECTION 1. THE BIJAGOS ARCHIPELAGO BIOSPHERE RESERVE (GUINEA BISSAU)

### 1.1. THE BIJAGOS ARCHIPELAGO

The Bijagos archipelago of Guinea Bissau is a natural wonder of global importance. It is an area of immense biological diversity and productivity, and contains a wide range of habitats from brackish water mangroves to beach-fringed tropical islands. In general the archipelago remains almost untouched by modern civilization and development, with little evidence of anthropogenic degradation or over-exploitation. The area provides important habitat for indigenous wildlife, migratory birds, and both resident and transient marine species. Some of these species are rare, unusual, or endangered. The archipelago also serves as the only home to the Bijagos people, whose culture appears as threatened as much of Africa's wildlife and natural habitats.

Not only is the Bijagos archipelago ecologically unique—it also provides an unparalleled opportunity to plan for development that is truly sustainable over the long-term. The abundance of natural resources and wildlife, the existence of a long-standing culture in equilibrium with its environment, and the pronounced interest of the Guinea Bissau government and foreign agencies and investors to develop the area wisely all bodes well for the future of the islands. The problem will be to decide what development, where, and for what objectives. Unchecked or poorly planned tourism development or fisheries activity could do irreparable damage to what is an essentially sensitive, carefully balanced ecosystem. The chance to protect the archipelago and its people will never be as significant as it is today: Guinea Bissau is poised a critical juncture regarding the future of a globally unique ecological treasure.

*Marine Protected Areas and Ocean Conservation*, by Tundi Spring Agardy.
© 1997 R.G. Landes Company.

Three basic steps must be taken to ensure that future development of the archipelago and the coastal zone of Guinea Bissau is undertaken in a sustainable way. The first step is ecological: making sure that the processes critical to the self sustainability of the system are protected. The second step is sociological: ensuring that local stewardship and long-term responsibility and commitment to conservation of the archipelago and its culture is maintained. The third step is political and economic: integrating a plan for the Bijagos Archipelago with long-term development activities for both the country of Guinea Bissau and the region as a whole. The latter point is an important one, for however well thought out a plan for the Bijagos islands, its ultimate success depends on the condition of the larger system; and no coastal plan exists in a vacuum.

The following case study discusses the concrete steps that have been taken to ensure that a viable, long-term coastal plan for the archipelago can be put into action. Such a plan must be compatible with the very real and crucial development needs of the country. It evaluates various proposals for protecting the archipelago, pinpoints geographical areas which ought to serve as foci for environmental protection, suggests where further research is most needed and how it can most effectively provide information necessary for resource management, and makes recommendations on how to build grass roots support and governmental commitment to long-term conservation. The major recommendations that follow concentrate on how to develop and implement a plan for a coastal biosphere reserve encompassing the archipelago. Other recommendations pertain to general sustainable use of marine resources and governance issues. But since conservation is in itself a dynamic process, planning must remain amenable to change over time to be effective.

The Bijagos Archipelago is located west of the Guinea Bissau coastline in West Africa. An extensive system of islands and coastal waters, it is comprised of 20 major inhabited islands, 26 smaller, transiently-inhabited islands, and some 37 uninhabited islets (Fig. 8.1). The continental shelf, on which the island chain rests, is the largest of any such area in West Africa and covers some 53,000 square kilometers. This shelf area, bathed by freshwater riverine input and strong offshore currents and upwelling, supports one of the most productive marine ecosystems in the region. The marine environs, in turn, support a large and diverse terrestrially-based suite of wildlife.

The archipelago was formed at the end of the last glaciation, when receding ice carved out the largest of the canals and separated the island chain from the mainland.[1] Since this relatively recent geological formation, the archipelago has undergone significant changes in form, with large scale shifts in location of sandbars and exposed mudflats. The actual coastlines are hard to distinguish because of the dynamic nature of these features.[2] Many of the numerous shell and silicate beaches

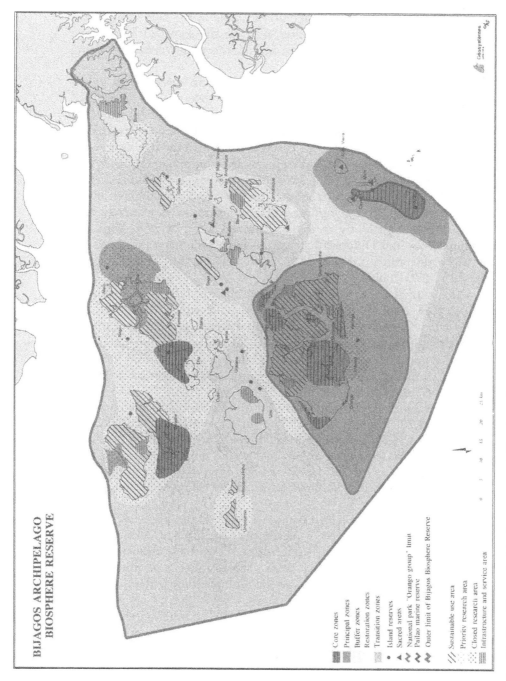

BIJAGOS ARCHIPELAGO
BIOSPHERE RESERVE

Core zones
Principal zones
Buffer zones
Restoration zones
Transition zones
Island reserves
Sacred areas
National park "Orango group" limit
Pailao marine reserve
Outer limit of Bijagos Biosphere Reserve

Sustainable use area
Priority research area
Closed research area
Infrastructure and service area

*Fig. 8.1. The Bijagos Archipelago proposed Biosphere Reserve.*

found throughout the archipelago are unstable in the face of high seas and currents.[3] Bodies of freshwater are rare: the so-called lakes being only transient accumulations of rain, and the island canals all being saline.

The microhabitat characteristics of the islands' coastal areas follow a gradual cline from brackish water, mudflat areas to rock and beach fringed offshore islands. The ecological character of the system is heavily influenced by alluvial output from the Geba and Cacheu Rivers, and by precipitation and associated coastal run-off patterns during the wet and dry seasons. Rainfall often reaches 2250 millimeters during the seven-month wet period from May to November. Humidity is generally high, averaging 50% throughout the year,[1] and reaching a monthly mean of close to 90% in summer.[3] The temperature is relatively constant, fluctuating slightly around the annual mean of 27°C with two maxima (May and October) and two minima (February and August).[3] Water temperatures average 24°C, with seasonal surface isoclines in the 23°C-27°C range.

The rainfall patterns and associated current flows are under the influence of seasonally variable yet predictable winds. There are five major winds, including: 1) the continental winds, bringing hot dry air from the northeast; 2) the trade winds, bringing cool and humid air from the north; 3) the monsoonal winds, bringing hot and very humid air from the southwest; 4) the grand equatorial wind from the east; and 5) the variable sea breezes.[3]

The archipelago contains over 150 kilometers of silicate and shell beaches, 94,200 hectares (ha) of intertidal rocky areas, 76,000 ha of mudflats, and 35,000 ha of mangrove forests.[3] The distribution of biomes is influenced by freshwater input and associated hydromorphological patterns, as well as by soil conditions and fragility.[1] Overall, approximately 57% of the soil is ferruginous, 32% halomorphic, and 11% marginal/rocky.[3]

Tidal activity is a major compoment of hydromorphology, with a tidal range of approximately 3 meters near the Cacheu delta and over 5 meters in the area of the Geba delta. Estuarine currents are also significant, ranging from 2.5-3 knots. The confluence of these coastal water flows and the strong ocean currents and upwelling patterns produces a highly dynamic system.

The major sea currents which bathe the archipelago during the dry season are called the "Canary waters;" in the wet season warmer waters from the south predominate.[2] Tidally-produced coastal currents are some of the strongest in the region. The sea temperature regimes, tidal fronts, and mixing all produce conditions ideal for photosynthesis and primary productivity.[2]

The phytoplankton production on the continental shelf waters provides the base for an extensive and stable marine food chain. This food chain, in turn, supports a diversity of terrestrial species, notably migratory and resident birds, as well as mammals and reptiles. Plant diversity on land is thought to be correspondingly high, with fragments of primary forests still evident on some islands in the archipelago.

The extremely phytoplankton-rich coastal waters of the archipelago form the base for a large and thriving marine trophic web. Populations of marine planktivorous species, upon which marine carnivores directly or indirectly depend, are large. Inshore, and especially in areas of brackish tidally mixed water, the dominant planktivores are mullet (*Mugil spp.*) and shrimp (*Penaeus notialis* and *Parapenaeopris atlantica*). Offshore, upwelling areas support a pelagic food chain which differs both in species and dynamics from the inshore food webs. Despite such differences, inshore and offshore systems are connected through both physical and biotic linkages.

Within archipelagic waters, the most common fish species are mullet (*Mugil cephalus*), shad (*Ilisha africana*), sea catfish (*Arius heudeloti*), barracudas (*Sphyraena sphyraena*), groupers (especially *Serranus aeneus*), snappers (particularly *Lutjanus agenes*), corvina (*Otolithus brachygnathus*), and numerous cartilaginous fish species, particularly of the genera *Raja*, *Dasyatis*, *Carcharhinus* and *Sphyrna*.

The actual composition and distribution of the ichthyofauna at specific locations is unknown, since the best available information comes from directed fisheries statistics. Inshore catch statistics from Guinea Bissau's coastal, non-artisanal fisheries show the majority of catch (by volume) is represented by estuarine sciaenids (55-62%), with other sciaenids, sparids, and crustaceans making up the bulk of the remainder in decreasing order.[2] Thus, mullets and sciaenids are the dominant species in the archipelagic coastal waters.

Molluscan species are common and form an important basis for fish and avian food chains (as well as being a major food source for humans). Mangrove oysters, arc shells (*Arca senilis*), and other pelycopods live in mud flat/mangrove areas. Gastropods, including volutes (*Cymbium pepo*), cones, cowries and conchs, are also abundant. Other invertebrate groups such as marine annelids are well-represented, and the meiofauna is thought to be rich in species diversity.

Marine, brackish water and terrestrial reptiles form an important part of the ecology of the archipelago region. Five sea turtle species have been recorded from Guinea Bissau waters, including the green turtle (*Chelonia mydas*), loggerhead (*Caretta caretta*), olive ridley (*Lepidochelys olivacea*), hawksbill (*Eretmochelys imbricata*) and leatherback (*Dermochelys coriacea*).[4] Freshwater turtles of the species *Pelusios subniger* inhabit creeks and wetlands areas. Two crocodile species are present in the archipelago: *Crocodylus niloticus* and *Osteolaemus tetraspis*. Six snake species are reportedly abundant (Methot, 1990) including *Python sebae, Boaedon fuliginosus, Psammophis sibilans, Dendroaspis viridis, Elapsoides semiannulata*.[3] Monitor and agamid lizards are also common, although systematic surveys of these reptilean groups are lacking.

The avifauna of the archipelago is very diverse, with estimates of 700,000 overwintering birds visiting the archipelago. Breeding pairs of resident species number 10-15,000.[5] The most common seabirds and shorebirds include terns, gulls, egrets, herons, spoonbills, ibis, cormorants, pelicans and ducks.[6] The archipelago has been hailed as a critically important migratory stopover and feeding area for migratory birds summering in Europe.

Marine mammal species are represented by the bottlenose dolphin (*Tursiops truncatus*), the humpbacked dolphin (*Sousa teuszii*), the manatee (*Trichachus senegalensis*), and, if one is liberal with the term "marine mammal," by the hippo (*Hippopotamus amphibius*). The Bijagos archipelago is one of the few places on the earth where the hippopotamus ventures into the sea, and they do so regularly. Four antelope species are known from the archipelago, including *Cephalophus monticola*, *Tragelaphus scriptus*, *Kobus kob* and *Redunca redunca*.[1] Primates are represented by two monkey species: *Cerocopithecus aethiops* and *C. petawis*. The Gambian rat, squirrel and several bat species comprise the remainder of the mammalian fauna.

The terrestrial habitats of the archipelago are in and of themselves diverse and variable, with lowland mangrove habitats in some coastal areas, savannas on the larger islands, and complex, successionally mature forests on a few of the islands. Approximately 96% of the surface area of the islands is forested, with 27% of this total palm forest, 27% herbaceous savanna and 31% mangrove areas.[3] The most common tree species, dominating the palm forests, is the palm oil palm (*Elaeis guineensis*).

## 1.2. RESOURCE USE PATTERNS

The human inhabitants of the Bijagos archipelago fall into three broadly-defined categories: 1) indigenous Bijagos people; 2) Nyominka fishermen originating in Senegal; and 3) immigrants from Bissau or elsewhere on the mainland and from foreign countries. The total population of the archipelago at last census (1979) was estimated to be 18,000 people living in some 140 villages.[3] Based on this census, the population density was determined to be 17 inhabitants per square kilometer, with a mean family size of 6.6. In 1979 the annual growth rate was estimated at 0.9%.

The resource use patterns of the indigenous Bijagos differ markedly from that of the immigrants. For the most part, the Bijagos are terrestrially-focused in their resource use, and have a largely sustainable hunter/gatherer and agrarian existence.

The principle utilization of natural resources for food is through itinerant rice culture, supplemented by oil of palm production.[1] Rice is the major staple of the entire country, and rice must be imported to supplement local production in the archipelago. According to Methot (1990) and other researchers, rice agriculture, particularly when practiced in non-mangrove areas, is appropriate as long as fallow field regeneration is respected and burning is controlled.[3] Palm production, on the other hand, is sufficient to supplement local needs and also supports a small export industry. Products from the palm tree (called "the tree of seven sustenances" by local peoples) include palm wine, palm fruit, wood for construction, thatch for roofing, fiber for clothing, and heart of palm, in addition to the palm oil itself.

Other crops grown on the Bijagos islands include peanuts, beans, peas, and medicinal herbs. All islands with permanent settlements have livestock, including poultry, cattle, pigs and goats, most of which are allowed to roam and forage freely.

Hunting among the Bijagos is practiced rarely, usually for religious ceremonies. Occasionally-hunted wildlife include manatees, sea turtles and crocodiles. Technically, hunting is forbidden in the archipelago, since it as a nationally-designated Wildlife Reserve falls under Article 9 of the "Regulamento de Caea."[3] Forest guards posted in the Reserve uphold such hunting regulations, but there are only four guards in the entire archipelago (in Bubaque, Uno, Orango and Formosa). Although little data are available, there is a notable absence of top terrestrial predators and large marine grazers. In the case of crocodiles and manatees this may be a function of overexploitation, since even low level hunting can significantly impact organisms with slow replacement rates.

Artisanal fishing, while practiced only rarely by the Bijagos, forms the basis of Nyominka life. Fishing is generally practiced close to shore, with pirouges or slender wooden canoes providing transport and gillnets and weirs as the standard gear. Mullet are the primary inshore species targeted, accounting for 90% of the artisanal fishery by volume (CEDR, 1989) and sharks and pelagic finfishes are the commonly targeted coastal species.[1] According to CEDR studies of the Boloma-based Nyominka fishery, artisanal fishermen average 10-15 trips per month, and yield an average of 150 kg of fish per trip in the dry season, and 7-8 trips per month and yield an average of 50 kg per trip in the rainy season. Fish are either sold to semi-industrial cooperatives for export from the archipelago, or are smoked, salted or dried for subsistence or local trade.[7]

Although commercial fishing pressures appear to be rising, development pressures generated from within the archipelago by residents have not increased precipitously. This is in part a reflection of demographics: absolute population growth is relatively low and emmigration of younger Bijagos (estimated to be approximately 20% since 1950) curtails the already low growth. The number of Nyominka fishermen fishing within archipelagic waters may have risen in recent years, although little information is available about artisanal effort. The growth of the transient population employed in the service industry in those few places that cater to visitors (Bubaque, Caravela, etc.) is noticeable but as yet probably has had little major impact on the ecosystem. On the other hand, if transportation to the islands is improved and if tourism and recreational fishing opportunities are maximized, the environmental health of the Bijagos archipelago may become seriously eroded from within.

The immensely rich waters of the Guinea Bissau coastal zone and Exclusive Economic Zone (EEZ) support an important and expanding large scale fishery as well. The commercial fishery, located primarily beyond the archipelagic waters, is large and forms a significant component of national revenue.[8] World Bank estimates suggest a country-wide maximum sustainable yield of 200,000 to 300,000 tonnes (of which 3-5,500 tonnes would be shrimp) at a potential (1986) market value of $310 million. Since the industrial fishing fleet is comprised entirely of foreign vessels licensed by the Guinea Bissau government, national revenues are generated only through licensing. Despite this, such licensing income accounted for almost 30% of total gross national product in 1988. It should be noted that since domestic involvement in the industrial fishing sector is nonexistent, the country has little incentive to regulate or restrict foreign fishing activities. This is especially true in light of the fact that little scientific evidence has been generated to show that overexploitation of Guinea Bissau's offshore stocks could negatively impact inshore resources and their use.

Major target species of commercial fishing operations include shrimp, squid, snapper, grouper, jacks, tunas and mackerel, billfishes and sharks.[7] Triggerfish (genus *Ballistes*) were once an important component of the industrial catch (over 30% of the total commercial catch at one time), but apparently overfishing has reduced stocks dramatically.[9] Trawling, seining and longlining are the most commonly employed fishing techniques.

### 1.3. EXISTING FRAMEWORKS FOR NATURAL RESOURCE MANAGEMENT

In an important sense, the archipelago's isolation and inaccessibility has made it a natural sanctuary.[10] This buffering has protected not only the resident wildlife, but also the culturally-unique Bijagos people. The isolation, both geographic and logistic, has certainly done more than any national, regional or local regulation to preserve the integrity of the island and its peoples. Yet the distance between the archipelago and the outside world decreases every day, and the luxury of de facto isolation will disappear all too soon.

Existing protection for the natural resources of the coastal zone of Guinea Bissau in general and the Bijagos archipelago in particular is only marginal. As mentioned previously, hunting of wildlife in the archipelago is forbidden by national law. Fishing is regulated to the extent that commercial fishing operations are excluded from territorial seas. Enforcement of such regulations is minimal and often ineffective, however.

A note must be made about sovereignty, stewardship and land rights in Guinea Bissau. All lands belong to the state, and, technically speaking, traditional use in no way confers rights of ownership.[3] The socialistic stance of the government about land ownership reflects a fundamentally egalitarian attitude about resources. Villagers are essentially concessionaires, although leasing is only formalized when development is commercial (e.g., the construction of a hotel or the operation of a restaurant).

Despite the "equal access" principle inherent in the property system, interviews with Bijagos villagers on some of the islands revealed that they had a strong sense of resource ownership: referring to "our sea turtles" or "our manatees," for example. The perception of resource ownership seems to generate a certain amount of friction, in that villagers resent the taking of "their" resources by outsiders.

This apparent competition for informally claimed resources is not uniform: in Orango, for instance, villagers in Ancopado voiced displeasure over the stealing of "their" sea turtles by people from the neighboring village, yet had no qualms about people from the big village to the east coming and taking those same turtles. This uneven attitude may be rooted in the traditional caste structure of the Bijagos, where the upper class people from the village of the king enjoy certain rights not granted to others in lower, subordinate classes. Such intrinsic regulation of resource use must be studied more carefully, as it could form the basis for stewardship-oriented management in the future.

Many departments of the Guinea Bissau government have an interest in resource exploitation and protection in the Bijagos even if they do not directly regulate activity there. The Ministries of Industrial Fisheries, Artisanal Fisheries, Tourism, Rural Development and Agriculture, Transport, and Natural Resources and Industry all have a role in the management of the archipelago. Local government has centers in Bubaque, Formosa and Umo, with regional administration in Bolama. Outside agencies and organizations with involvement in resource management or protection in the archipelago include Solidami, the United Nations Development Program, the United Nation's Food and Agricultural Organization, the World Bank, numerous aid agencies, AFVP, CECI and IUCN, among others.[10]

National sovereignty over marine areas follows the Law of the Sea Treaty model, with waters within the area framed by the coastal baseline to 12 miles offshore defined as territorial seas, and waters 12-200 miles offshore defined as the Exclusive Economic Zone (EEZ).

## 1.4. COASTAL MANAGEMENT NEEDS

There is no paucity of interest in the preservation of the Bijagos Archipelago, yet the sensitive balance of this productive and diverse ecosystem stands to be disrupted by poorly planned development. The only thing that will ensure that the health of the archipelago and its inhabitants remains strong, and that fishing, tourism and other development remain sustainable over the long-term, is planning that is comprehensive, ecologically sound, and cognizant of real and prospective environmental threats. Ultimately, the plans to protect critical habitats and processes in the archipelago will have to be incorporated into a national coastal management plan to make them truly effective.

The objectives of the conservation work undertaken by the author in the early 1990s were to evaluate existing plans to protect the archipelago's environment, including a long-standing proposal to designate the archipelago as a biosphere reserve. To critique this and other proposed conservation actions, basic questions about the nature of the ecosystem and its present condition, as well as the nature of anticipated future threats, needed to be addressed. These specific questions were considered:

1)   How is the archipelagic ecosystem threatened at present?–Are current artisanal fishing practices too destructive or overly intensive for the system to support?–What about commercial fishing and its impact on the inshore ecology?–How do agricultural patterns effect the health of the environment (including cutting or burning of vegetation, and the maintenance of stock animals)?

2)   How do "upstream" activities such as nutrient-loading, deforestation and subsequent erosion, pollution, etc. affect the ecosystem of the islands and environs?

3)   Does or will tourist development conflict with fishing or other traditional uses?–Will the sensitive environment of the coastal areas be able to withstand increased tourism activity?

4)   How does the political situation regarding land ownership and traditional patterns of land and sea use affect the sense of stewardship over resources and the perception of people's rights to exploit them?

5)   What are the physical features of the environment that influence productivity?–In which geographic areas and when are the biotic processes which drive the ecosystem concentrated?

6)   Why have certain areas been targeted for protection?–What is the nature of anticipated regulated uses around those protected areas?

7)   How realistic is it to assume that areas can be effectively protected?–By whom, and at whose cost?

8)   Can local individuals or organizations be identified to carry the concepts of integrated resource management beyond merely the designation of a protected area?–How will wise development plans be implemented over the long-term?

Cursory answers to most of these questions were derived from on-site missions, through literature reviews and interviews with government personnel, researchers and inhabitants of the archipelago. Many of these questions, however, require further attention and may need to be addressed following implementation of the initial biosphere reserve plan.

## 1.5. CRITICAL AREAS FOR CONSERVATION

Terrestrial areas in the Bijagos Archipelago that are of outstanding ecological importance have already been targeted for protected area designation in areas of high biodiversity, high productivity and good environmental condition. Yet because the linkages between the land and sea are so strong in this archipelagic environment, it is fundamentally important to protect critical marine areas as well as land-based core areas. The archipelagic ecosystem of the Bijagos islands is dominated by marine processes, and linkages between offshore, inshore and terrestrial habitats are complex and well-developed. The high biological productivity of the system as a whole rests on the enormous phytoplankton production of nearshore waters, and on the mixing and nutrient delivery patterns produced by the confluence of oceanographic and alluvial hydrodynamics. The physical environment of the coastal zone influences all the terrestrial habitats and acts as a buffer, playing a homeostatic role in ecosystem dynamics. Biologically, the extensive marine food web supports a diverse terrestrial fauna, including significant numbers of invertebrate and vertebrate species.

Based on the limited available scientific information about living and non-living resources in the archipelago, several types of critical habitats can be identified. In terms of marine primary production, brackish water shallows formed by the convergence of mudflats and creeks seem to be critically important. Such areas support the greatest densities of invertebrates and planktivorous fish, providing an enormous base to the trophic pyramid. Mangrove areas, particularly in the larger inland canals, are also important in terms of production, and support a unique suite of fish fauna and rare vertebrates such as the manatee, crocodiles, and even sea turtles.

Areas of freshwater loading are also important, since many species are adapted specifically to fresh or brackish water environments. The condition of the aquifers and the alluvial flows which provide freshwater to the coastal system must be maintained to support the great biodiversity of the region. The amount of freshwater loading to the system is also important, as fish and other marine production may be influenced by what is known as "rainfall variation stress."[11]

The canals between the islands or associated tidal flats are also critical to ecosystem dynamics, providing a physical mechanism for the delivery of coastal nutrients offshore and the delivery of oxygen-rich offshore waters to inshore areas. In addition, canals provide corridors

for migratory and semi-migratory fish, and may be important in reproduction and recruitment.

On the islands themselves, wide open silicate beaches which are
relatively stable are critically important to those species which nest on
or in the sand. Such species include the four species of endangered
sea turtles, as well as numerous species of seabirds.

Offshore foraging areas are also critical for sea turtle species. Since
herbivorous green turtles form the largest proportion of the total sea
turtle population, a habitat of primary interest to their protection is
the seagrass bed. Seagrass beds are critically important as feeding areas
for herbivorous fishes, mollusks and crustaceans as well, and are an
important spawning ground for many fish species.

Farther offshore, structurally heterogeneous areas such as rocky reefs
provide important habitat and spawning grounds for many pelagic fish
species. These rocky areas may also be important in primary production, with macroalgae providing an important food resource for invertebrate and vertebrate grazers. Lastly, rocky marine areas provide an
important nursery habitat for vast numbers of fish and crustacean species.

## 1.6. RESOURCE USE CONFLICTS

Given the importance of the commercial fishery to the national
economy and the creeping infringement of fishing activities on coastal
areas supporting other uses (tourism, etc.), fishing is quite possibly
the most threatening form of coastal development in Guinea Bissau.
Although the fishing industry has not reached the prominence that it
has in Senegal (Fontana and Weber, 1982), the trend is for expansion
of the whole spectrum of fisheries, from low level subsistence fishing
to large scale industrial trawling and seining.[12] One great danger is
that the expansion of both commercial and artisanal fishing activities
will proceed uncontrolled and without the incorporation of fisheries
management into general coastal planning.[13,14] Another is that in the
rush to expand the sector and reap its economic benefits, fishing methods
will be adopted that are culturally or environmentally inappropriate.[15]

The scale of this threat, that is, of the impact of inappropriate or
poorly planned fisheries expansion, is difficult to assess in Guinea Bissau.
An important clue can be gleaned from talking to the fishermen themselves, yet most sociological studies in the region concentrate on the
Senegalese fishermen. Although it is true that much of the artisanal
fishing is practiced by Nyomikas from Senegal, generalizations cannot
be made from surveys done on such fishermen in Senegal (e.g., see
Fontana and Weber, 1982).[12] While it is tempting to say that Nyominka
fishermen have similar taboos and attitudes about the social status of
fishermen, the claim may be invalid. For instance, a study done by
Bacle and Cecil (1989) points to a potentially favorable relationship
between fisheries and tourism in Senegal: the majority of fishermen
interviewed felt there was no reason to feel threatened by potential

conflict, since tourists tend to eat lots of fish![16] Whether this perceived harmony extends to Guinea Bissau, where fishermen seem generally reclusive and less interested in profit-making, remains to be seen.

Nonetheless, the potential for competition and conflict between the growing artisanal and commercial fisheries is increasing all the time. According to Bernacsek (1987), this conflict is already serious in many parts of Africa.[11] He states that "widespread lack of management and control results in conflicts between small-scale and large-scale fisheries. In Africa, cases of direct interference are increasing, including trawlers running over and destroying gear, trawling in nearshore waters reserved for artisanal fishermen, even deaths. Eventually competition for the same stocks will make itself felt by falling catches in the small-scale fishery." It should be noted that early warning signs are already present in Guinea Bissau: four foreign vessels were apprehended within Guinea Bissau territorial waters in 1989 according to the World Bank Guinea Bissau Office, and the Nyominka fishermen interviewed stated average sizes of both mullet and some carnivorous fishes have decreased in recent years.

Although direct competition may be a factor in the above declines, it is more likely that the negative impacts of both artisanal and industrial fishing on inshore stocks has occurred through indirect interference. Though the artisanal fishery is small in scale and thus relatively unintensive, gear is set in important corridors for migration and may interfere with reproduction and/or recruitment. Similarly, offshore industrial fishing may impact the marine food web and the linkages between the inshore and offshore systems.

Other resource use conflicts which are beginning to emerge in the archipelago have to do with the partitioning of land/sea areas for traditional use versus non-traditional activities such as tourism. To date these competing activities have not been mutually exclusive, and there is considerable ground for expanding the ecotourism market without necessary negative impacts (see below).

## 1.7. ANTICIPATED GROWTH SECTORS AND POTENTIAL IMPACTS

Major growth in three sectors can be expected in the coming years: fisheries (both commercial and semi-industrial), tourism, and transportation. The degree to which any of these industries will develop, and the rate of expansion, cannot be predicted. However, it is clear that the government of Guinea Bissau has little economic incentive to check burgeoning development in any of these sectors, and without such checks we can expect that growth in all of these sectors will be maximized.

Future fisheries development may be problematic, since many assessments already point to stressed, and in some cases, overfished stocks.[16] Non-selective gear such as is employed by most of the commercial fishery is bound to effect not only the target stocks but others as well.

Large scale fishing activity may well be undermining the trophic structure of the marine system, bringing the entire coastal system closer to the point of collapse. Furthermore, incidental catch of species already threatened by low population sizes (such as sea turtles) may cause significant ecological damage.

Even the growth of the semi-industrial fishery may present environmental problems if it is not carefully planned and monitored.[17] The current proposal to place a large-scale processing platform in the vicinity of Pailao suggests that this careful planning is lacking. There is some international interest in assessing the impacts of such semi-industrial and industrial fishing activities on the health of the coastal ecosystem (e.g., see refs. 18-21), but whether these voices are being heard remains to be seen.[18-21]

Expansion of the currently low-level tourism market may have implications for the environment as well. Besides the ecological impact of constructing infrastructures to support such an expanding industry, increased human activity in sensitive nearshore environments may well stress the system. Limited resources such as freshwater might also be exploited beyond the natural carrying capacity of the islands.

Tourism in the Bijagos Archipelago does not necessarily have to be costly to the environment or local culture, however. If properly balanced, the socio-economic benefits of tourism can include local employment, local economic gains, better social welfare, the generation of foreign exchange, and better services to inhabitants of the archipelago.[22] According to Budowski (1976), while "unplanned or poorly planned tourism in areas of conservation significance can lead to conflict, a change of attitude leading to a symbiotic relationship between tourism and conservation can provide cultural, ethical and economic benefits."[23]

Such is the nature of eco-tourism, in which protected natural habitats and indigenous cultures provide the objects of tourism interest. Eco-tourism is defined as "that segment of tourism that involves travelling to relatively undisturbed natural areas with the specific objective of admiring, studying and enjoying the scenery and its plants and animals, as well as cultural features."[22] Eco-tourism in the Bijagos Archipelago could provide significant revenues without necessarily causing corollary environmental degradation. Achieving the right balance between small-scale eco-tourism which does not provide appreciable local revenue and large-scale eco-tourism which taxes the environment may be difficult, however.[24]

Although the Bijagos Archipelago is currently unequipped to deal with major growth in the tourism sector, some growth is inevitable. Tourism is an enormous business. In 1989, 420 million travellers spent U.S. $40 billion globally. Adventure travel, including eco-tourism, is the fastest growing segment of the tourism industry.[22]

Tourism in general and eco-tourism specifically will likely focus in the following areas of the archipelago: Caravela, with its 10 kilometers of pristine beaches and relatively clear coastal water; Uhnocomo with its recreational fishing potential; Canhabaque with its birdlife and beaches; and perhaps Carache, Uno, and Orango with their diverse habitats.[20] There are currently five factors which limit both eco-tourism and conventional tourism growth: 1) the accessibility of the islands; 2) infrastructural deficiencies; 3) an inadequate agricultural base for supplying local foods for tourists and service employees; 4) under-developed sanitation and security; and 5) limited ecological carrying capacity.[1] The latter term is nebulous and was not described by the authors, but presumably includes limited freshwater supply. All these factors, plus the specific sensitivities of land and coastal habitats accommodating tourists, must be weighed against the prospective benefits of expansion.

The line between tourism which is beneficial to the local communities and the nation and tourism which negatively affects the archipelago is hard to decipher. It is thus absolutely essential that the administrators and planners of tourism in Guinea Bissau communicate with managers of all other activities. This communication can foster a symbiotic relationship: one in which the Tourism Ministry benefits from the information on resources provided by other ministries and organizations, and the latter benefit from being able to determine what future activities to incorporate into their assessments. The lack of such fundamental communication is a recipe for disaster.

## 1.8. THE BIOSPHERE RESERVE NOMINATION

The opportunities to develop an integrated plan to manage the resources of the archipelago and protect traditional uses of those resources are enormous, given the strong national and international interest in the area. The most effective way to amalgamate these diverse interests and ensure that future development of archipelagic resources does not compromise either the environment or local peoples is through zoning. Such zoning must be developed to protect ecologically critical processes and habitats, through protected conservation cores and regulated use buffer areas.[25] This is the principle behind UNESCO's Biosphere Reserve Programme (Batisse, 1990), and the reason that the concept is being adapted for use in this particular area.[26]

In order to develop a biosphere reserve plan that effectively protects the environment in the face of strong anticipated development pressures, the following questions must be rigorously addressed:

1) What are the objectives that the reserve and its zoning plan sets out to fulfill? Are these objectives the same for administrators (local and national government, etc.) as they are for the inhabitants of the archipelago?

2)   What are the physico-biotic processes which are critical to maintaining the ecosystem?

3)   How can core areas be designated to protect these critical processes?

4)   What uses must be regulated in areas surrounding the cores to ensure they are buffered from external negative impacts?

5)   How can an effective biosphere reserve plan be carried out? How can the local public become involved, in order to develop a strong sense of stewardship and long-term support? What administrative mechanisms can be put in place to facilitate implementation of the reserve plan?

The answers to some of these questions are discussed below; others will have to be answered when new data are collected and as new information emerges.

Delineating the outer bounds of a prospective biosphere reserve requires consideration of three factors: the ecology of the system and its natural bounds, the political organization underlying resource management and its geographical representation, and the degree to which the bounded area can be realistically managed. Ideally, all three of these boundaries should coincide–but they rarely do. Deciding on the actual outer bounds of the biosphere reserve entails tailoring the system as it is defined ecologically into a workable geographic area.

Much discussion has been undertaken about the limits of a Bijagos biosphere reserve. From a systems standpoint, such a biosphere reserve should include a major portion of the Geba and Cacheu deltas, as well as Boloma and some of the other inshore islands. From an administrative viewpoint the inclusion of Boloma also makes sense, since the administrative seat of regional government is there. Yet for practical reasons, the consensus among conservationists is that the biosphere reserve should only include the ecologically-similar outer islands of the archipelago (Fig. 3.1).

Core areas, if properly designed, act to protect ecologically critical processes on land and in the sea.[27] Theoretically, development can be sustained even in light of increased levels of exploitation if and only if the ability of the system to keep itself going and replace its depleted resources is not impaired. Thus core areas become an investment in the future: a way of capitalizing on the interest without diminishing the capital. The more is known about the dynamics and functioning of an ecosystem, the less this process of identifying critical core areas is burdened with risk.

Our knowledge of the coastal system of Guinea Bissau is far from complete, yet incomplete knowledge is not a valid excuse for abandoning the cause for ecologically-based management (despite the fact that it is usually cited as one). Clearly, the dominant features of this system are water-based, with both ocean and riverine processes contributing to its dynamics. Thus it is important that marine cores, in addition to terrestrial ones, are identified and protected.

Several terrestrial conservation cores have been identified by researchers;[18,28] these include: 1) a vast core area encompassing the island of Imbone in the Orango group; 2) the island of Poilao and associated islets; 3) a small inland portion of Canabaque; and 4) the northern section of Caravela Island. These cores are noteworthy because of their relatively pristine condition, their high diversity, and the presence of special species of interest such as sea turtles and hippopotamus.[29] Such cores are analogous to "sites of special scientific interest" used in coastal planning in Great Britain and elsewhere.[30]

It is clear that protecting these few areas will not be sufficient to protect the system as a whole and ensure that future development can be sustained.[31] Additional core areas need to be designated in areas important for marine primary production, fish feeding, fish spawning, nursery functions and critical migratory corridors.[32] Preliminary indications about where those critical areas are located exists, however, further information is needed to identify marine cores.

Additional cores may need to be established to protect special rare species such as sea turtles.[33] The core area in Poilao (Limoges, 1991) goes far towards achieving this, but the seasonal variation in the boundaries of that conservation core are insufficient and unrealistic.[4] To design a conservation area which will really protect the local populations of sea turtles, one needs to target the nesting beaches, foraging areas and migratory corridors used by these animals.[34] Further research will be necessary to fine-tune this and other conservation cores.

The design and implementation of buffer areas around conservation cores is as least as important as the appropriate designation of the cores themselves. The purpose of buffer areas is to shield the cores from potentially adverse impacts of human activities outside the core. In the fluid marine environment, such buffers are especially important.[27] However, management and mitigation of impacts is more difficult in marine and coastal systems, so the buffer zones must be planned with an eye to realism. In other words, designing an elaborate system of regulated use areas with little prospect of being able to control those uses does virtually no good.[35]

It takes an accurate and thorough assessment of real and prospective patterns of resource exploitation to be able to design buffer areas. The need for ministries involved in this development to communicate among themselves is, as stated previously, critical to this process. Although much of the future development potential of the Bijagos archipelago cannot now be determined with certainty, certain restrictions on development can be proposed to reduce the risk of undermining the core areas and the general health of the ecosystem.

For terrestrial core areas, buffering occurs as a function of being able to fence off large enough fragments of the habitats to be self-sustaining. Thus terrestrial buffers can be designed with general uses in mind, and only minor restrictions. People can inhabit buffer areas and use the resources within them to the maximum extent the carrying capacity will allow. One important provision for land-based buffer zones should be the requirement of environmental impact assessments (EIAs) for all commercial development activities which might impact nearby core areas. Most important would be the assessment of prospective tourism developments. Mossman (1987) provides some guidance in this, suggesting that those in charge of tourism development be required to: 1) determine the area's carrying capacity for tourism; 2) ensure that tourist developments are appropriate to the natural and cultural environment; and 3) ensure that the scale of development is compatible with the primary values of the protected area.[36]

For example, the core area on Caravela should be surrounded by a buffer area which encompasses the island areas on either side of the core. Although regulations within this buffer would not restrict low level exploitation of natural resources for subsistence, they would require any tourism developers planning facilities in the buffer zone to prepare EIAs showing all potential environmental impacts on the neighboring core. Were this to be adopted within the coastal management plan for Guinea Bissau, the government could require such EIAs and have the costs of such assessments borne by the developers themselves. Such EIA requirements would proffer Guinea Bissau with the added benefit of generating more information about the ecology of the area at no cost to the country.

Marine buffer areas are slightly more problematic.[37] Given the nature of threats to the marine environment in Guinea Bissau, most of the regulations in marine buffer areas ought to be directed at fishing: industrial, artisanal, and recreational. Seasonal restrictions could be developed near identified spawning areas to effectively safeguard the stocks. In areas used as migration corridors, restrictions on fixed gear fishing (gill netting, for example) would be useful. A complete restriction on fishing in feeding areas would pertain only to cores, with additional restrictions on pathways used by fish, sea turtles and the like to get to feeding grounds.

Increased boating and shipping activity related to tourism and other development should also be a major consideration in the design of buffers. Restrictions on motor boat traffic in areas where manatees and sea turtles congregate is one such regulated use. In these cases, however, the prospect for enforcing such regulations must be taken into account. Without local interest in protecting resources, such regulations may be for naught.

Kelleher and Kenchington (1989) provide guidelines for the planning of marine protected areas which should be cited here.[38] They claim that the following elements help to guarantee the effectiveness of a protected area: 1) simplicity; 2) a minimum of regulation; 3) maximum consistency with existing laws; 4) use of buffers to make the transition between protected and non-protected areas; and 5) identification of discreet geographical characteristics as the foundation for zoning patterns. In an ideal world, the entire archipelago would be a buffer zone: a carefully-controlled area of regulated multiple uses. But in a less than ideal world, where controlling use is problematic, buffers must be designed to be large enough to include truly critical areas but small enough to manage effectively.

## 1.9. LESSONS LEARNED FROM OTHER RESERVES AND PROTECTED AREAS

Valuable lessons can be learned from the experiences (both positive and negative) of others who have worked to develop coastal biosphere reserves and marine protected areas. To date, over 500 marine protected areas greater than 1000 hectares in size have been designated around the world.[39] A scant few of these provide good examples of integrated marine resource management. Those examples with relevance to Guinea Bissau's conservation plans come from the Great Barrier Reef Marine Park in Australia, the Virgin Islands Biosphere Reserve, plans to protect Samana Bay in the Dominican Republic, the Galapagos Marine Park and the Saba Marine Park in the Netherlands Antilles.

The Australian example provided by the Great Barrier Reef Marine Park has its value in demonstrating the concept of multiple use zoning.[40] However, there are important differences between Guinea Bissau and Australia: Australia's barrier reef is immense (over 1400 kilometers long), highly productive, and only marginally threatened by human activity. The pattern of zoning established by the Park Authority relied heavily on the existing pattern of uses, which were largely segregated and non-conflicting. Lastly, the pressure to develop and exploit Australia's coastal resources is nowhere near as great as that of Guinea Bissau.

The scale of the Virgin Islands Biosphere Reserve is more comparable to that of the Bijagos Archipelago, but the nature of resource use conflicts and economic considerations is vastly different. The Virgin Islands Biosphere Reserve is a multiple use marine and terrestrial protected area that was superimposed on the existing Virgin Islands National Park. Although the National Park occupies some two-thirds of the land area of St. John in the Virgin Islands, it became obvious to park researchers and managers that protecting the island alone was not sufficient. Thus the Biosphere Reserve was designed to encompass nearshore waters around the island, as well as island communities and land that fall outside of park boundaries.[41]

This expansion of protection has been very successful in maintaining the health and value of the insular system. Although no economic analysis has been done, it is assumed that an intact and well-managed ecosystem has greater appeal to eco-tourists visiting the area (which number in the hundreds of thousands). Thus the protection of the ecosystem has indirectly supported economic development, as has provided other benefits to local peoples.

The ongoing efforts to protect Samana Bay in the Dominican Republic provide several lessons for biosphere reserve planning. Although the Bay is not yet designated as a biosphere reserve, great strides have been made in protecting both the land and sea areas there. In planning a reserve at Samana, the focus for conservation and sustainable development has slowly expanded to include more and more local residents and interest groups. The greatest value of this case study is in demonstrating how to involve local peoples in the planning process.[42]

The Galapagos Islands of Ecuador, although famous for being unique, nonetheless provide interesting parallels to the Bijagos Archipelago. Both island systems support a high diversity of terrestrial and marine species, have extremely productive nearshore waters, have rare species of special interest, and support a local human population heavily dependent on the environment. The importance of eco-tourism to the economy, clearly demonstrated in the Galapagos and potentially significant in the Bijagos, is a central feature of planning.[43] Yet the protection of the marine environment in the Galapagos came almost as an afterthought to a long-standing land-based conservation movement–despite the fact that the systems are inseparable.[44] Guinea Bissau is showing considerably more foresight in planning marine protection in concert with terrestrial conservation work.

Saba Island's Marine Park provides an excellent example of how small-scale, community-based conservation can be effective and good for the local economy. As discussed in chapter 5, Saba is a small island in the Netherlands Antilles with a spectacular reef around it. Recognizing the value of this reef to a potentially important eco-tourist (scuba diving) industry, the government decided to protect all coastal waters as a marine park. Although well protected, the park does accommodate a variety of potentially conflicting uses without conflict, made possible by a well-planned multiple use zoning system.[25] Today Saba supports a thriving tourist industry, which contributes to its local economy not only through direct revenue-generation but also through the marine park user fee system.

These examples show how a well protected and healthy coastal and marine ecosystem can support a variety of uses in a sustainable and economically viable way. The precise mechanism for achieving this balance and making a multiple use protected area truly work differs according to local conditions. One of the most valuable features of the coastal biosphere reserve concept is its inherent flexibility, so the planning process can readily be fine-tuned to meet local needs.[45]

## 1.10. INSTITUTIONAL INFRASTRUCTURE AND COOPERATION

It is one thing to design an integrated protected area; it is quite another to implement the plan successfully. Kelleher and Kenchington (1990) provide some useful guidelines on implementation, with a strong plea for simplicity in planning. For the Bijagos Biosphere Reserve, an

important key to success will be the appropriate integration of the
reserve with already existing activities.[38] A new bureaucracy to implement the reserve is probably unnecessary; instead, management of the
area should be made a clear and prominent priority for already established
offices in the federal government of Guinea Bissau.

Having said this, the importance of having local involvement in
the administration of the archipelagic reserve cannot be understated.
Hence a two-tiered management structure may be most appropriate:
federal coordination and local on-the-ground management.

The specific infrastructure needed to implement and manage the
reserve will have to be determined as the project evolves. One possibility is that federal administration of the reserve be housed in the
Ministry of Agriculture and Rural Development, with a multi-minis-
terial advisory committee involving the ministries of Tourism, Fisheries, etc. Since fiscal support for the reserve is likely to come from
international agencies, their representation on the committee is also
advisable. For the sake of simplicity, a single person within the Ministry of Rural Development and Agriculture should be identified as
the federal manager of the reserve, with only this responsibility.

For local administration of the reserve, most effective would be
the establishment of a Bijagos Council, comprised of local residents,
aid and development agency representatives, and members of local
government. The structure of this council could be modelled on the
example provided by the Sian Ka'an Biosphere Reserve in Mexico. The
Council should be established to provide a democratic means of voic-
ing concern, with equitable voting rights. A Council president could
then communicate directly with the federal reserve manager, under a
formal agreement that local decisions would have a strong representation
in the federal management of the reserve.

## 1.11. GUIDELINES FOR PLANNING AND IMPLEMENTATION
## OF THE BIJAGOS RESERVE

At a minimum the following steps should be undertaken to ensure
the development of an effective and beneficial biosphere reserve in the
Bijagos Archipelago:
1.  Help focus efforts on the development of a country-wide
    coastal management plan, of which the Bijagos conserva-
    tion work is an important and illuminating model.
2.  As part of the above process, help catalyze the efforts to
    establish a multi-ministerial Environmental Commission, to
    facilitate communications between government agencies and
    ensure that sustainable use philosophies have adequate
    representation.
3.  Support efforts to manage Orango effectively as a national
    park, but only as a part of integrated coastal management
    efforts (in other words, be careful that the park designation

does not give the mistaken impression that conservation interests have been adequately served).

4.  Tap the wealth of information on marine resources that the Nyominka fishermen possess through one-to-one interviews. The person/persons visiting Nyominka camps should be careful to establish a good rapport with the inhabitants, to lay the groundwork for their later involvement in the planning and implementation of a multiple use protected area.

5.  Devote significant time and resources to build alliances with residents of the archipelago, including such influential people as local clergy and the President of Guinea Bissau, as well as villagers and hotel operators.

6.  Continue archipelago-wide marine surveys to identify critical areas.

7.  Facilitate, to the maximum extent possible, the efforts to collect and analyze fisheries data, particularly that information which bears on the relationship between offshore and inshore processes.

8.  Help launch a stranding/incidental by-catch reporting network in the fisheries observer program, to shed light in the ecosystem-wide impacts of commercial fishing in the EEZ.

9.  Develop a means to include an EIA requirement for tourism and other development in the coastal zone.

10. Ensure that the ever-growing international efforts to develop or protect the Bijagos archipelago are carried out on a harmonious and synergistic way, by encouraging monthly multi-agency/organization status meetings and frequent communication. In addition, it will be important to continue to look for examples of how conservation benefits development, track the success (or lack thereof) of marine protected area case studies, and communicate those findings to the government and development banks, in an ongoing effort to build a strong conservation ethic in Guinea Bissau.

The work to conserve the natural resources of the Bijagos archipelago could well serve as a globally-important model for sustainable development. It could equally as well demonstrate how enormous amounts of money and human energy were wasted in the blind rush to exploit. The difference in outcome will reflect what questions are addressed in the next few months, and how well potentially conflicting uses can be planned for in an integrative, forward-thinking way. The potential for success in the Bijagos is in and of itself enormous incentive to continue striding forward.

## SECTION 2. THE MAFIA ISLAND MARINE PARK (TANZANIA)

### 2.1. MAFIA ISLAND ENVIRONS

Mafia Island is located in the western Indian Ocean just south of the island of Zanzibar and southeast of the Tanzanian capital, Dar es Salaam. The marine environment is famous for having an extraordinary diversity of marine species, and Mafia Island's reefs are some of the least disturbed on the east African coast. Coral, sponge, and reef fish species are plentiful, making the area notable as a "hot-spot" for biodiversity. The island's ecosystems are also a critical seed bank, providing food, shelter, and breeding grounds for seabirds, waders, and migratory birds, sea turtles, and marine mammals. The local economy is highly dependent on the island's rich fisheries.

Many of the coastal and marine habitats in the greater Mafia Island region form critical habitats for dugong (*Dugong dugon*) and sea turtle species (the green turtle *Chelonia mydas*, the hawksbill turtle *Eretmochelys imbricata*, the olive ridley *Lepidochelys olivacea* and the loggerhead sea turtle *Caretta caretta*).

The coral reefs of Mafia Island are among the richest in the world.[46] Much of this diversity is due to the fact that the island sits in close proximity to the nutrient-rich Rufigi River delta area of the Tanzanian mainland. These productive mangrove and shallow water habitats provide nutrition and living spaces for many of the species that live on the reef, for part if not all of their lifetimes. Yet the silted freshwater from the Rufigi River is kept from lowering the water quality around the reefs thanks to the strong northward flow of the East African Coastal Current, which diverts sediments away from the coralline habitats. Thus the extensive coral reef and mangrove systems have been able to develop in optimum conditions due to the island's prime location between the nutrient-rich waters of the Rufigi Delta and the dynamic continental shelf system.

The cave and grotto formations in the exceptionally old and extensive fringing reefs of Mafia Island provide a diverse array of microhabitats, colonized by a diversity of plant and invertebrate species. The area's coral reefs, mangroves, and softbottom communities support over 380 species of fish, 45 genera of sceleractinian corals, 12 species of seagrasses, 7 mangrove species, over 134 species of algae and some 140 species of sponge.[47] The inhabitants of Mafia Island and its associated islets depend on this richness in marine resources for their livelihoods. Until recently, these communities lived in near equilibrium conditions with their environment, using marine resources sustainably and maintaining a high quality of life.

The fisheries that the Mafia Island ecosystems support are known throughout the east African region as being highly productive. These fisheries are important to Mafia Islanders for subsistence protein as well as income through local markets and sales to regional traders. Both the men and women residents of Mafia Island are fishers and catch octopus, finfish, sea cucumbers and lobster. The inhabitants of Mafia Island also collect or farm seaweed using traditional methods. However, overexploitation of some species and destructive fishing practices have added to environmental stresses already caused by coral mining, pollution from terrestrial sources, anchor damage, deforestation of mangroves, ship-borne pollution, and other impacts arising from continued population rise.

Unfortunately, as in many other coastal regions worldwide, the Mafia Island region is slowly becoming degraded. The northern portion of the island's coastal waters has witnessed extensive overfishing. Harmful practices such as dynamite fishing, practiced largely by outsiders, and sand mining have destroyed extensive coastal and benthic habitats. Mafia Islanders fear that as fisheries are depleted farther north, the nearshore waters of the island will attract even more exploitation.

The marine systems of Mafia Island are under ever-increasing threat from overfishing, destructive and non-sustainable exploitation and indirect environmental degradation. Although dynamite fishing was the original catalyst that spurred local people to demand protection of the reefs (see below), other resource use is also unsustainable. In the past, Mafia Islanders utilized dead coral rubble for lime production, which forms the basis for local construction. As reserves of rubble were used up, however, the local communities were increasingly forced to mine living coral to augment lime derived from rubble. This mining, combined with ever-increasing fishing pressure, has interfered with critical marine processes in many reef systems. Loss of biodiversity and accelerated rates of erosion are only two of the many signs that this form of environmental degradation threatens the Mafia Island system. Poor land use practices have exacerbated stresses on the coastal system, and Mafia's high productivity and diversity now stand threatened.

## 2.2. CORAL REEF CONSERVATION: SPECIAL CONSIDERATIONS

The biologically rich and diverse coastal ecosystems of Mafia Island are threatened by ecologically-damaging use of marine resources and by degradation from afar. These threats, in turn, undermine the ability of Mafia Island's residents to live sustainably and in harmony with their island environment. To abate this destruction, scientific guidance on the design, implementation and monitoring of a marine park encompassing Mafia's most important and productive coastal habitats was needed.

Coral reef conservation issues epitomize the difficulties we have in protecting the marine environment as a whole. Reefs are well known as biologically rich, interesting systems. Vibrant in color, form and species, and a virtual catalogue of specialization, the coral reefs compete with tropical forests for our attention and concern. Unfortunately, we seem to be loving them to death. Coral reef communities are extremely sensitive to environmental change and disturbance, and our fishing, scuba diving, glass bottom boating and swimming has begun to take its toll on reefs the world over. This, combined with geographically large scale effects like pollution and increasing temperatures that result in coral bleaching, has acted to reduce genetic and species diversity before our very eyes.

There is a sense of urgency for catalyzing effective action for coral reef conservation. Most human communities in the coastal tropics exhibit an enormous dependency on intact, productive reefs and the resources they offer. As this dependency grows and the pressures to harvest resources and use reefs increases, cumulative impacts are undermining the health of reef communities and overall degradation threatens even some of the most remote reef areas.

Coral reefs, formed by the deposition of the limestone skeletons of thousands of tiny, anemone-like animals that live symbiotically with the tiny coral algae in their tissues, are the mega-diversity areas of the oceans. Corals form three distinctive types of structures: fringing reefs close to the shore, barrier reefs separated from the mainland by lagoons, and atolls that are circular reefs around islands which have long since submerged.

For conservation and management purposes, it is important to distinguish between hermatypic (or stony) corals, and ahermatypic (non-symbiotic) corals. The true reef-building corals are the hermatypic polyps with symbiotic algae (zooxanthellae) within their bodies. The algae process the coral's wastes, thereby recycling vital nutrients to make a significant contribution to the high productivity of the system. A coral reef is defined as a population of stony corals which continues to build on products of its own making. Reefs are the characteristic coral formations in the most tropical oceans. However, in some areas like West Africa and parts of Southeast Asia, non-reef building coral communities are more likely to occur. These are assemblages of stony corals, often in deeper, colder waters, that do not build on themselves, or ahermatypic corals which cannot build reefs. Coral communities may rival coral reefs in high productivity, and merit as much conservation attention as the reefs themselves.

Coral reefs are crucial for local fisheries, in Mafia Island and around in the world. They provide the food fish, molluscs, and crustaceans for many coastal communities, and function as breeding grounds for many commercial species upon which even inland peoples depend. They are especially important for maintaining traditional fisheries in island nations. In addition, reef-related tourism can be a very important source of foreign currency. And given the extraordinary biodiversity of reef ecosystems, it is not surprising that an ever-increasing number of reef species are found to have medicinal properties.

Coral reefs are fragile ecosystems in the sense that regrowth is slow in comparison to the rate of damage.[48] Reefs grow no more than 12 meters (and often much less) in 1000 years. Most reef communities are dynamic and are adapted to natural disturbances, which contribute to the maintenance of the system's high diversity. Nevertheless, they do have rather strict environmental requirements. Coral colonies need light, oxygen, water temperatures between 22°C and 28°C, and low loads of suspended sediments.

This sensitivity spells doom for many coral reefs.[49] They can be killed by predation (e.g., by crown-of-thorns starfish), by smothering with sediments or algal overgrowth, and by bleaching from high water temperatures. These biologically diverse and productive ecosystems are threatened the world over by poor land use resulting in erosion, nutrient overloading from sewage pollution and agricultural run-off, coral mining for use as building materials, dynamite fishing, careless tourists and tour operators, coastal development and global climate change.

Resource extraction on the reefs is also a major, devastating problem. Overfishing of pufferfish, triggerfish and other reefal species can indirectly damage reefs by causing outbreaks of their prey, sea urchins, which graze on algal turf, thereby eroding the coral rock surface and weakening the reef structure.[50] Similarly, overfishing of algal-grazing fishes can trigger the spread of harmful algae on the surface of the corals, causing widespread degradation by smothering. Even more insidious, the use of destructive gear such as pole nets and dynamite, or poisons such as chlorine and cyanide, can cause widespread and lasting damage–sometimes irreversibly.[51]

Worldwide, one of the most serious threats to corals is from sediment runoff, caused by deforestation, slash-and-burn agriculture, and loss of mangroves, which act as sediment traps. Pollution from sewage originating from coastal settlements causes eutrophication, stimulating the growth of algae and smothering corals.[52] Coastal agricultural or agrarian development, or agriculture bordering watersheds, has a similar over-fertilizing effect on coral reef systems. And though not all reef systems are true coral reefs (some reefs are structures, such as algal ridges, formed from calcifying red algae), even these non-coral reefs are vulnerable to human-induced damage, caused by changes in environmental conditions that alter the growth rate of calcareous algae themselves or disrupt balances in the ecology of the ecosystem.

In order to conserve coral reefs and ensure that they will continue to provide the goods and services that coastal communities so highly value, it is necessary to identify those parts of the coral reef ecosystem that need the most stringent protection.[53] Areas of reef that are highly productive or harbor a wide diversity of species are obvious targets for conservation. However, there may be additional areas that need to be protected in order to keep the reef ecosystem thriving. Such areas include soft-bottom communities in adjacent areas, seagrass meadows that serve as feeding and breeding areas for many reef organisms, mangrove areas that provide nutrients and nursery areas for many species, and the major migration corridors that link these diverse critical habitats. Any coral reef protection program will have to conserve all these critical areas to some degree in order to be effective.

### 2.3. HISTORICAL BACKGROUND

Mafia Island residents highly value the marine ecosystem and the resources it provides, and recognize the threats now facing this area. For this reason, local community leaders asked the Tanzania government and prominent non-governmental organizations to assist in the development of a plan to protect Mafia Island's most vital elements. Support for a protected area plan was overwhelming, although by the time the park was officially legislated, many residents shared a sense of frustration about the slow pace of the process.

The establishment of a marine park at Mafia Island had been under discussion since 1968, when marine biologists from the United States first noted the globally important biodiversity of the island's coral reefs. Two small reserves had been previously established in the Mafia Island region, but their management proved ineffective because of the paucity of human and financial resources, and the lack of a substantive legislative base.

In 1988, the islanders, representatives of the Tanzanian government and conservation organizations met to discuss ways to help local people protect the resources upon which their livelihoods depend. They proposed creating a large, multiple use marine reserve. The first step in the process was a series of on-island workshops to allow islanders to frame objectives and express their expectations and desires. During the workshops, the residents of Mafia Island made it clear that they wanted strong action to protect marine and coastal resources of the insular ecosystem. Non-governmental organizations then worked with scientists to help local representatives draft a park management plan. Implementing this plan and identifying innovative financing mechanisms for its management, as well as developing micro-enterprise projects to remove intensive pressures on marine resources are all ongoing.

Mafia Island residents worked with technical experts to come up with a multiple use zoning plan. Three categories of protection within the reserve's boundaries were developed: one strict protection and regulations preventing the removal of any natural resources; a second with regulations restricting the amount and type of fishing; and a third permitting general use with a minimum of controls (Tables 8.1 and 8.2). Dynamite fishing and other inherently destructive practices are not allowed anywhere within the park's boundaries.

A major objective for the establishment of a marine protected area, in addition to the goals relating to curbing destructive fishing practices, was to allow residents of Mafia Island and surrounding communities

### Table 8.1. Mafia Island Marine Park general zones and regulations

**Mafia Island marine park zoning regulations**

| Uses | General development Reg. zone | Cons. zone | Core zone |
|------|------|------|------|
| Tourism: | | | |
| sightseeing | * | * | * P1 |
| swim & snorkeling | * | * | * P1 |
| diving | P | P | P1 |
| fishing | P | X | X |
| Scientific research: | | | |
| seaweed studies | *P | P1 | X |
| fish studies | *P | X | X |
| crustacean studies | *P | X | X |
| mollusc studies | *P | X | X |
| Oil & gas exp. or recovery | X | X | X |
| Waterfront development | P | P | X |
| Tourism infrastructures | P | X | X |
| Drainage/sewerage | P | X | X |
| Waste disposal | X | X | X |
| Sand extraction | X | X | X |
| Industrial development | X | X | X |
| Seabed mining | X | X | X |
| Dynamiting | X | X | X |
| Beach sand mining | X | X | X |
| Shipping | P1 | X | X |
| Seaplane landing | P1 | X | X |
| Aquarium fish collecting | P | X | X |
| Mangrove exploitation | P | X | X |
| Marina development | P | X | X |
| Non-motorized boating | * | * | * |
| Jet skiing | X | X | X |
| Water skiing | X | X | X |
| Parasailing | X | X | X |
| Horse or camel riding | P | X | X |
| Collection of non-comm. spp. | P | X | X |

**KEY:**  * = activities fully allowed      P = allowable by general permit
P1 = allowable by resident permit
P2 = allowable by non-resident permit   X = prohibited

to exert some control over tourism development on the island. Mafia Islanders are Islamic peoples with certain cultural sensitivities regarding clothing and behavior that tourism threatened to undermine. In other East African islands such as Lamu to the north of Mafia, tourism has essentially run amuck and the cultural beliefs of the local people have been largely destroyed by conflicting value systems of tourists.

*Table 8.2. Mafia Island Marine Park fishing zones and regulations*

**Mafia Island Marine Park zoning regulations**

| Uses | Fisheries Reg. Zone | Cons. Zone | Core Zone |
|---|---|---|---|
| **Fishing by residents:** | | | |
| handlining | *P | P1 | X |
| box trapping | *P | P1 | X |
| shark netting | *P | X | X |
| seining | *P | X | X |
| fence trapping | *P | P1 | X |
| octopus collecting | *P | P1 | X |
| sea cucumbering | *P | P1 | X |
| lobster/crabbing | *P | P1 | X |
| live shelling | *P | X | X |
| coral harvesting | P | X | X |
| **Fishing by non-residents** | | | |
| handlining | *P/P2 | X | X |
| box trapping | P/P2 | X | X |
| shark netting | X | X | X |
| seining | P/P2 | X | X |
| fence trapping | P/P2 | X | X |
| octopus collecting | X/P2 | X | X |
| sea cucumbering | X/P2 | X | X |
| lobster/crabbing | X/P2 | X | X |
| live shelling | X | X | X |
| coral harvesting | X | X | X |

**KEY:** * = activities fully allowed     P = allowable by general permit
P1 = allowable by resident permit
P2 = allowable by non-resident permit   X = prohibited

Thus the Mafia Island Marine Park would have to address questions of culturally and environmentally sensitive future development.

Interestingly, local residents were adamant about having large areas protected as strictly protected cores within the multiple use reserve. This was the case even though residents knew that it entailed restricting their own activities. Mafia Islanders have become staunch advocates for conservation, and put strong pressure on the government of Tanzania to pass the Marine Reserves Act of Parliament (1994) that made creation of Mafia Island Marine Park possible. They similarly campaigned for the passage of the Act of Parliament that created the Mafia Island Marine Park (1995). Although Tanzania currently has 25% of its land area included within a network of parks and reserves, the country had no marine parks until the Mafia Island Marine Park was established.

## 2.4. Mafia Island Marine Park

The Mafia Island Marine Park of Tanzania was designed as a multiple use, zoned marine protected area. Unlike the Bijagos Archipelago Biosphere Reserve, the initial boundaries of the park encompass only part of the coastal ecosystem (the southern coast and waters of the island). The small size of the park is a reflection of political reality: Tanzanians felt it prudent to start small, with the eventual intent of scaling up to have the reserve cover the entire insular ecosystem.

Substantive steps towards the establishment of a multiple use protected area, initially encompassing the waters and coastal lands of the southern portion of Mafia Island, were taken in an October 1991 workshop. At this workshop, held on Mafia Island with representatives of all local communities in attendance, agreement was reached on the geographical scope of the protected area, its general objectives, basic zoning plan (Fig. 8.2) and administrative infrastructure. World Wildlife Fund and other non-governmental organizations assisted with the arrangement and funding of this and subsequent workshops. A further workshop on the zoning of Chole Bay was carried out with the six villages in the area, including a boat trip to produce agreed demarcation maps. This workshop was very successful, and similar workshops will be held for zoning other areas of the park. Village meetings have been particularly useful in bringing the communities into the zoning process, and each of the twelve villages within the park boundaries has formed a committee to liaise with the park authorities on management activities.

The crusade against dynamite fishing was the first substantive conservation activity to get underway, largely due to the strongly motivated support from the village representative and the full District Council. The village representatives were angry about the heightened levels of explosive fishing, and were more than willing to assist in efforts to stop it. As a first step, a task force called the Central Coordinating Committee (CCC), directly responsible to the District Commissioner, was created. The role of the CCC is to formulate policy and to provide an independent mechanism by which the actions of the Department of Natural Resources authorities involved in the arrest, prosecution, and conviction of offenders can be evaluated. The Committee has links with the villages, who act as the surveillance mechanism for "Operation Dynamite." Sail powered fishing boats known as dhows are used as undercover enforcement patrol boats—discouraging would-be dynamite fishermen from setting off explosives as long as any dhow is in the vicinity (which, in these heavily used waters, is all the time).

The management plan for the entire Mafia Island Marine Park has also been drafted, but will need to be revised as more information

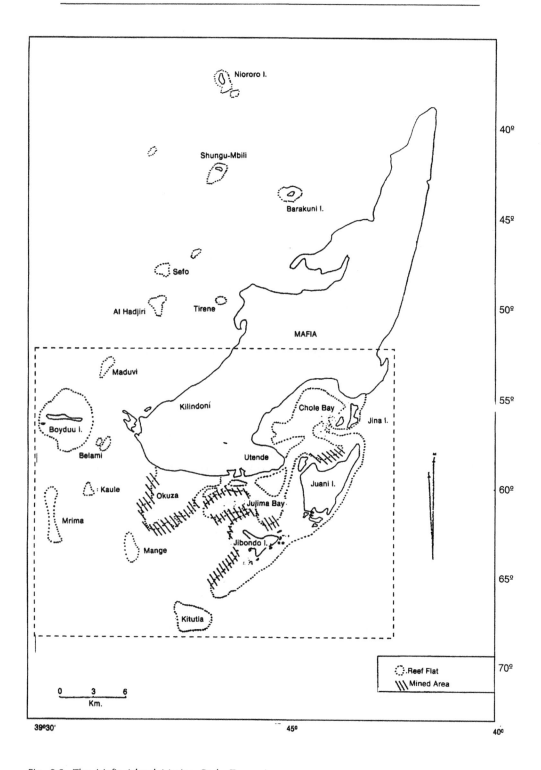

Fig. 8.2. The Mafia Island Marine Park, Tanzania.

needed to fine-tune the zoning plan is obtained. The process is guided by a committed group of people from the region: the Technical Committee presently consists of the Tanzanian Fisheries Division, Institute of Marine Science of the University of Dar es Salaam, Member of Parliament for Mafia Island, the Wildlife Conservation Society of Tanzania and the World Wildlife Fund Country Office of Tanzania. Additional expertise has been provided by Kenyan marine biologists and park planners, several socio-economists and other conservationists. Work is progressing to institute a mechanism for the development of management plans for other similar marine parks.

Mafia Island residents are now undertaking enforcement patrols and have been trained to be able to perform scientific monitoring to ensure that conservation objectives are indeed being met. This monitoring is twofold: biological monitoring to evaluate how the ecosystem is responding to management and protection and social monitoring to assess whether people's needs and expectations are being sufficiently met. The precise methodology for monitoring must reflect the very specific objectives that each of the park zones are meant to target. For example, core areas that aim to protect community structure and thus maintain high levels of biodiversity, protect habitats from physical destruction, maintain levels of productivity, and provide vital habitat for critical processes will be monitored through periodic transects to determine percent coral cover, fish community structure and abundance counts, surveys to check for disease and physical alterations to the site, etc. Further work will be needed to finish designing the monitoring methodology, to develop a social monitoring system in parallel to the biological one, and to ensure that the cost/benefit ratio of the monitoring plan is minimized.

Three tasks must be completed for the successful implementation of the Mafia Island Marine Park: 1) fine-tuning the Mafia Island Marine Park Management Plan, including the design of a carefully-constructed monitoring and evaluation component to measure biological and social efficacy of management; 2) baseline studies to amend the zoning plan, if necessary, and provide a basis for evaluating whether or not project goals are being attained; and 3) planning and initiating several small scale community development projects to continue to build the confidence of local people and demonstrate the benefits of their involvement in conservation. These elements include an effective management plan, baseline information, an ongoing evaluation methodology, and community development projects, which will not only guarantee the long-term success of the MIMP, but also pave the way for other marine conservation efforts.

Given the growing frustration that local communities have had with the slow pace of progress in getting the park legislated (four years from the first workshop to the passage of the Act of Parliament), it is especially important to promote and support corollary projects on the ground. It is imperative to the development of stewardship and local pride in the park that several small scale community development projects are initiated. Such projects include support for the boat-building operation in Jibondo, extension support for the women's octopus fishing cooperative, and a project to identify alternative materials for construction (to ease pressure to harvest coral for lime) and begin training in use of these other materials.

The success of the park will in large part depend on how careful the development management plan was, and how flexible it will turn out to be–so that improvements to the management of Mafia Island's resources can be made as knowledge is acquired and needs change. To this end, the Technical Committee has done a stellar preliminary job.

The Mafia Island Marine Park is a notable marine protected area for a number of reasons: 1) Mafia Island Marine Park is Tanzania's first multiple use marine protected area, and as such will lay the groundwork for a future network of national marine parks, 2) the protected area targets some of the most highly diverse and most productive ecosystems in the Indian Ocean, and 3) Mafia Marine Park is being viewed as a model for marine protected area design and community-based management, not just for African coastal areas but in fact for similar efforts around the world. Expectations and stakes are thus high, and the success of Mafia Island's Marine Park will have wide-ranging repercussions.

## REFERENCES

1. Commission d'Etat du Developpement Rural (CEDR). Development integre de la zone IV: region de Bolama. Rapport de la Phase I: Etudes et propositons preliminaires. CRAD & SUCO report, 1989.

2. Domain, F. Rapport des campagnes de Chalutages du N.O. Andre Nizery an large des cotes de Guinee-Bissau. Institut de Recherche Agronomique de Guinee, Ministre Francais de la Cooperation, 1988.

3. Methot, S. Etude preliminaire de l'archipel des Bijagos en vue de la creation d'une aire protegee. CECI publication, 1990.

4. Limoges, B. Preliminary report on sea turtles in the Bijagos archipelago. Unpublished report, IUCN Bissau, Guinea Bissau, 1991.

5. Campredon, P. The Bijagos archipelago. Case study for the Workshop on Coastal Biosphere Reserves, 14-20 Aug 1989, San Francisco, CA.

6. Naurois, R. Peuplements et cycles de reproduction des oiseux de la cote occidentale d'afrique, du Cap Barbas, Sahara, Espagnol, a la frontiere de la Republique de Guinee. IVieme partie: Guinee Portugaise. Mem. Mus. Hist. Nat., Ser. A, Zoologie 1969; 56: 191-251.

7. da Silva, F. and J.L. Kromer. Projet d'amelioration des techniques artisanales de transformation du poisson. Projet du development integre des Iles Bijagos. Report 1988; 635/85/W02.

8. Bage, H.E., J.M. Kassimo, K. Steen, T.A. Vaz and I. Tvedten. Proposition de projet la peche dans l'archipel Bijagos, Guinee-Bissau. Project report, IUCN Bissau Office, 1989.

9. Fonds Africain de Development. Rapport d'evaluation: projet de developpement de la peche artisanal avancee, Guinee-Bissau. Dept. de l'Agriculture et du Developpement Rural II, 1990.

10. Lemay, M. Projet d'appui a la gestion integree des resources marines de l'archipel des Bijagos dans le cadre de l'establissement d'une reserve de la biosphere. CEIO, Division de l'afrique de l'Ouest et de l'Ocean Indien, 1990.

11. Bernacsek, G.M. Status of the fisheries sector in Africa. Rome, Food and Agricultural Organization, Fisheries Dept., Fishery and Policy Planning Division Report, 1987.

12. Fontana, A. and J. Weber. Apercu de la situation de la peche maritime Senegalaise. Centre de recherches oceanographiques de Dakar-Thiaoye. Dakar, Senegal, 1982.

13. Johnson, J.P. and M.P. Wilkie. Pour un development integre des peches artisanales: du bon usage de la participation et de la planification. Field Guide 1. FAO/DIPA, 1986.

14. Programme and Strategy Review Committee. African fisheries: reversing the decline. Report of the first meeting of the PSRC of the IDRC Fisheries Programme for Africa and the Middle East, 20-21 Mar 1990, Nairobi, Keyna. PSRC 1: The Nairobi Consultation.

15. Cycon, D.E. Managing fisheries in developing nations: a plea for appropriate development. Nat. Res. J. 1986;26: 1-14.

16. Bacle, J. and R. Cecil. Artisanal fisheries in Africa: Survey and research. CIDA. Hull, Canada, 1989.

17. Iles de Paix. Proposition de projet la peche dans l'archipel Bijagos Guinee-Bissau. ASBL, 1989 Report.

18. International Center for Ocean Development (ICOD). Project proposal: cooperative initiatives for integrated marine resources management in the archipelago of the Bijagos. Unpublished report, 1990.

19. International Union for the Conservation of Nature and Natural Resources (IUCN). Conservation du millieu et utilisation durable des resources naturelles dans la zone cotiere de la Guinee-Bissau. Rapport d'Activite Decembre 1989-Novembre 1990, 1990.

20. International Union for the Conservation of Nature and Natural Resources (IUCN). Planificao costeira da Guine-Bissau. Proposta Peliminar. Bissau, Guinee-Bissau, 1990.

21. International Union for the Conservation of Nature and Natural Resources (IUCN). Conservacao e desenvolvimento da zona costeira, Guine-Bissau. Bissau, Guinee-Bissau, 1990.

22. Ceballos-Lascurain, H. Tourism, ecotourism, and protected areas. Paper presented to the Commission on National Parks and Protected Areas (CNPPA), World Conservation Union (IUCN) General Assembly Meeting, 28 Nov–6 Dec, 1990, Perth, Australia.

23. Budowski, G. Tourism and environment conservation: conflict, coexistence or symbiosis? Env. & Cons. 1976; 3(1):27-31.

24. Boo, E. Ecotourism: Potentials and Pitfalls. World Wildlife Fund, Wash., DC, 1990.

25. Agardy, T. Advances in marine conservation: the role of protected areas. Trends in Ecology and Evolution 1994; 9(7):2676-270.

26. Batisse, M. Development and implementation of the biosphere reserve concept and its applicability to coastal regions. Envir. Cons. 1990; 17(2):111-116

27. Kenchington, R.A. and M.T. Agardy. Achieving marine conservation through biosphere reserve planning. Env. Cons. 1990; 17(1):39-44.

28. DGFC/CECI/IUCN. Projet de zone de conservation des tortues marines de l'archipel des Bijagos. Projet planification cotiere, 1991.

29. Robillard, M.J. and B. Limoges. Proposal for coastal planning. Unpublished document, CECI, Bissau, 1990.

30. Gubbay, S. Using sites of special scientific interest to conserve seashores for their marine biological interest. A report for the World Wide Fund for Nature from the Marine Conservation Society, London, England, 1989.

31. Diegues, A.C. Application of the biosphere reserve concept to coastal and marine areas: Case Study #8. Proceedings of the Workshop on Coastal Biosphere Reserves, 14-20 Aug 1989, San Francisco, CA, UNESCO.

32. Hatcher, B.G., R.E. Johannes, and A.I. Robertson. Review of research relevant to the conservation of shallow tropical marine ecosystems. Oceanogr. Mar. Biol. Ann. Rev. 1989; 27:337-414.

33. Sepa-Marie, M. Profit a court terme on exploration durable des tortues marines de l'archipel des Bijagos. Unpublished document, CECI, 1990.

34. Agardy, T. Last voyage of the ancient mariner? BBC Wildlife Dec. 1992:30-37.

35. Kelleher, G. Political and social dynamics for establishing marine protected areas. Key paper, Workshop on Coastal Biosphere Reserve, 14-20 Aug 1989, San Francisco, CA. UNESCO, 1989.

36. Mossman, R. Managing protected areas in the South Pacific: A training manual. SPREP/IUCN/NPS International Affairs joint publication, 1987.

37. Batisse, M. Development and implementation of the biosphere reserve concept in coastal areas. Proc. of the Workshop on Coastal Biosphere Reserves, 14-20 August 1989, San Francisco, CA. Paris, UNESCO, 1989.

38. Kelleher, G. and R. Kenchington. Political and social dynamics for establishing marine protected areas. Proc. of the UNESCO/IUCN Workshop on the Application of the Biosphere Reserve Concept to Coastal Areas, San Francisco, CA, 1989.

39. Kelleher, G., C. Bleakley and S. Wells. A Global Representative System of Marine Protected Areas. Washington, DC, World Bank, 1995(4 volumes).

40. Kelleher, G. Identification of the Great Barrier Reef region as a particularly sensitive area. In: Proc. of the International Seminar of the Protection of Sensitive Sea Areas, 1990: 170-179.

41. Towle, E.L. and C.S. Rogers. Case study on the Virgin Islands Biosphere Reserve. Case study, Workshop on Coastal Biosphere Reserves, 14-20 Aug 1989, San Francisco, CA. UNESCO.

42. Ferreras, J. Protected coastal areas: Samana Bay. Proceedings of a Workshop on management and planning of protected tropical coastal areas, Santo Domingo, Dominican Republic, 1987.

43. Broadus, J.M. A special marine reserve for the Galapagos Islands. Proceedings of Coastal Zone'85, 1985.

44. Broadus, J.M. and A.G. Gaines. Coastal and marine area management in the Galapagos Islands. Coastal Management 1987;15:75-88.

45. Agardy, T. The Science of Conservation in the Coastal Zone: New Insights on How to Design, Implement and Monitor Marine Protected Areas. Proc. of the World Parks Congress 8-21 Feb. 1992, Caracas, Venezuela. IUCN, Gland, Switzerland. 1995

46. International Union for the Conservation of Nature and Natural Resources (IUCN). Biodiversity in sub-Saharan Africa and its islands: Conservation, management and sustainable use. IUCN Species Survival Commission Occasional Paper 1990;6. Gland, Switzerland.

47. Gaudian, G. and M. Richmond. Mafia Island Marine Park Project. The People's Trust for Endangered Species. London, Imperial College, London, 1990.

48. Kenchington, R.A. and E.T. Hudson. Coral Reef Management Handbook. Paris, France. UNESCO, 1984.

49. Craik, W., R. Kenchington and G. Kelleher. Coral reef management. In: Dubinsky, ed. Ecosystems of the World. 1990 Vol. 25: 453-466.

50. Bohnsack, J.A. The potential of marine fishery reserves for reef fish management in the U.S. Southern Atlantic. NOAA Tech. Mem/NMFS-SEFC-261, 1990.

51. Johannes, R.E. and M. Riepen. Environmental, economic and social implications of the live reef fish trade in Asia and the western Pacific. Arlington, VA, The Nature Conservancy Report, 1995.

52. Hutchings, P.A. Biological destruction of coral reefs: A review. Coral Reefs 1986;4(4): 239-252.

53. Bakus, G.J. The selection and management of coral reef preserves. Ocean Management 1983; 8:305-316.

# PART IV:
# THE GENERAL PROCESS FOR
# PLANNING AND IMPLEMENTING
# MARINE PROTECTED AREAS

# CHAPTER 9

# MARINE PROTECTED AREA SITE SELECTION

## SECTION 1. SELECTING SITES BY VALUE, THREAT AND OPPORTUNITY

Until recently, most marine protected areas around the world were established merely because the opportunity for creating a marine park or reserve presented itself. Such opportunities were and are created when government, a special interest group or the local community sees the need to either protect a threatened marine habitat or wants to develop an area in a sustainable fashion. Thus the siting of most marine protected areas is a function not of an objective analysis of marine conservation needs, but rather an ad hoc process through which decision-makers take action when and where they can.

There are two main driving forces for initiating marine and coastal protected area planning: opportunities and threats. From a positive viewpoint, marine conservation projects can and should be supported in biological rich, diverse, or ecologically important areas and in areas where their demonstration value can be maximized. On the other hand, conservation is most urgently needed where poorly planned development, overuse of resources, or indirect degradation threaten to undermine the very resource base on which coastal peoples depend.[1] Evaluating threats and opportunities in a systematic way can provide a means to direct marine conservation support so that it is maximally effective and long-lasting.

Government agencies, regional bodies, and local decision-makers need to think strategically about three questions concerning marine conservation work. The first question is *what* needs doing? (i.e., What are the main marine conservation issues that can and should be addressed through marine protected areas?). The second question is *where* should it be done? (i.e. Where is the magnitude of threat to marine ecosystems and the degree of opportunity such that investment in marine

*Marine Protected Areas and Ocean Conservation*, by Tundi Spring Agardy.
© 1997 R.G. Landes Company.

protected areas is justified?) And the third question is *how* should it be done? (i.e., What techniques—scientific, sociological, political—can and should be harnessed to reach our goals?).

In terms of the *what*, we have already discussed priority issues that warrant attention (chapters 1 and 2). These issues include: 1) preserving ecologically critical habitat to ensure that resources may be used in a sustainable fashion; 2) analyzing the problem of eutrophication and other land-based sources of degradation to show how it can be quantified, dealt with, used as a focus for a public campaign, and how its effects both hinder conservation and exacerbate other problems like toxics; and 3) controlling inherently destructive fisheries, especially bottom trawling and dynamite/poison fisheries.

The *where* is a much more difficult question, discussed at length in the following pages. Identifying the specific geographical areas where marine protected areas might be established is something best done on a national or regional basis—however the process by which high priority areas might be identified is universal. Criteria used to pinpoint specific areas would include estimates of ecological importance (degree of endemism, species richness, productivity and degree to which important ecological processes are concentrated in an area) as well as quantification of direct and indirect threat.

Finally, the *how* question is one that should be answered on a case-by-case basis in the regions. General models of marine protected areas can provide some guidance in elaborating useful methods of marine conservation, but the appropriateness and feasibility of any particular design is something that only the local people or their representatives can judge. There is no question, however, that lessons learned have great value to all programs, and an exchange of this nature can only help us be more effective in tackling marine conservation problems.

There is no shortage of problems facing coastal areas today, nor difficulties in identifying geographical areas where protected areas plans might be relied upon as a tool to help solve some of those problems. In fact, the scope of the coastal conservation issue is so overwhelming that decision-makers and managers often don't know where to begin. Science can help narrow the scope of potential opportunities, and pinpoint areas where marine protected area implementation will be most likely to achieve conservation and sustainable use objectives.[2]

To the extent that there is a "formula" for identifying science-based areas of interest, this formula consistently relies on three major considerations: 1) ecological value, defined in terms of ecological richness (biodiversity and/or productivity, uniqueness of habitat or community, representativeness, or ecological services that the area provides); 2) degree of threat that directed exploitation or indirect anthropogenic impacts are exerting on the system; and 3) degree of opportunity (involving an assessment of feasibility and the extent to which local communities, governments and regional authorities can and want to conserve their environment through protected area planning. Broadly speaking, the tripartite analysis thus centers on ecology, resource use and impact, and opportunities for success.

A cursory glance at the history of marine parks and protected areas reveals that more often than not, only one of the three central elements to area identification is generally considered when planners choose a site. In the United States and other developed countries, site selection is often guided by opportunity: an area popular for its aesthetic or recreational value, largely untroubled by pollution and other forms of degradation, largely without resource use conflicts–is deemed a good opportunity for successful implementation of a park plan.[3] In other parts of the world, marine protected areas emerge as a response to threat, with little investigation into the potential for success and desires of local communities. And in places where ecological richness has provided the starting point for identifying areas of interest, simplistic measures involving species accounting have often been the sole measure of "value" of an area for marine park designation.

The first step to be undertaken in any sort of site selection process is a specific elaboration of what objectives a network of marine protected areas is being established to reach. If the overall objective of marine protected area planning is to conserve ecologically critical areas that are most highly threatened by direct and indirect impacts of human activity, the following methodology is recommended to identify appropriate sites for marine protected areas:

1.  Assess the state of knowledge about ecosystems in the region;
2.  Identify ecologically important areas using multiple criteria, such as high species diversity, high degree of endemism, significant productivity, concentration of important ecological processes, habitat for keystone or commercially important species, high habitat diversity (beta diversity), etc.;
3.  Assess threats to high priority coastal and marine areas identified above;
4.  Assess the degree to which existing management measures counter threats; and
5.  Identify opportunities for successful marine protected area implementation.

Species diversity is the foundation for what has recently become known as the "hot-spots" technique for area identification. Although this lamentable technique is most often used in prioritizing geographical areas of interest, it has recently been adapted for coastal and marine use. However, high diversity areas may not be true priority areas for protection if the objective of protection is to safeguard critical processes that keep ecosystems functioning in a balanced, productive fashion.[4] Although significantly high diversity areas may merit attention because they reflect the conditions necessary for maintaining large numbers of competing species in sympatry, they may not be most critical to the marine ecosystem as a whole.[5] This is particularly true in marine systems, where pockets of endemism are rare and habitats are functionally linked over wide distances. Some relatively species-poor marine areas, such as saltmarshes, mangroves, seagrass beds, etc., may be more important because their productivity supports enormous food webs. Such an apparent contradiction is exacerbated by the fact that species accounting tends to center on vertebrates or other easily identifiable taxa, when in fact functional and species diversity is best illustrated by microbes.

How, then, is it possible to produce a broad-brush overview of a region in order to identify which areas may be more important than others? Ideally, scientific information on system ecology, utilizing physical and biological oceanography and hydrology, can provide answers on where critical ecological processes are concentrated when. Even rudimentary information on currents, mixing, nutrient loading and bottom topography can provide salient information on critical areas. Such areas of interest can be mapped conventionally or using Geographic Information Systems (GIS) technology in order to provide the baseline for area identification.[6,7]

When scientific information concerning bathymetry, currents, and other basic features of the environment are patchy or difficult to obtain, indicator species or other surrogates can be used to determine the importance of an area. The distribution and abundance of keystone species, if they have been identified as such, often provide basic understanding of ecosystem function. In some areas where scientific research is sporadic or non-existent, data on use of commercially or culturally important species may be used as a rough approximation of where some critical areas can be found (e.g., see Papua New Guinea Conservation Needs Assessment; Agardy and Pernetta, 1992).[8] However, when using such surrogates to identify priority areas for protected area work, it is important that physical features of the environment be coupled as well as possible with information on the distribution of such "special" species.

In summary, the following is a five step process for identifying areas of interest where marine protected areas might be established:

Step 1. Identify critical features of the environment, distribution of ecologically and/or economically important species, and, if possible, location of concentrations of critical ecological processes—and map this information using conventional cartography or GIS.

Step 2. Assess levels of resource use and identify sources of degradation in the region, mapping resource use conflicts, directed threats (e.g., areas of overexploitation), and indirect threats (e.g., sources of pollution, destruction or degradation of functionally-linked habitats).

Step 3. Determine geographical areas where opportunities for conservation may be maximized (e.g., areas where local communities are predisposed to accepting the notion of protected area planning, areas that are well-studied, areas identified as priority areas by decision-makers).

Step 4. Superimpose maps containing information outlined in steps 1-3 (using ecological information as the base map, overlaid by threats, and finally overlaid by windows of opportunity) to see where priority areas fall out.

Step 5. Investigate the feasibility of protected area planning in the priority regions, and begin the planning process by defining, with the strong involvement of local users, the specific objectives that such a protected area may help to achieve.

## SECTION 2. DEFINING THE NATURAL BOUNDS OF THE TARGETED ECOSYSTEM

The decision to initiate a coastal conservation project in a specific locale is a value-laden one. However, scientific information can help assess the feasibility of doing integrated coastal management in one area relative to others. Science is also important for defining the gross unit of management, addressing questions such as (inter alia): How big an area should the management encompass? How far upstream? How far out to sea? Should entire islands or just their coastlines and nearshore areas be included?[9] Fully three kinds of criteria are important for focusing the allegorical conservation spotlight: 1) physical linkages involving currents, freshwater hydrology, weather, source–sink models; 2) biological linkages involving important predator–prey relationships, migration, breeding and spawning movements, factors affecting productivity (Chapin et al., 1992); and 3) sociological linkages, including cultural ties between communities, political or management infrastructure and others.[10]

Wherever possible, coastal marine protected areas should try and influence land use in watersheds that impact the nearshore through nutrient and pollutant loading, hydrology, wetlands connections, etc. The extent to which a watershed is included in the protected area itself is a question of feasibility—most estuarine or coastal protected areas near rivers will likely only include the river mouth. However, a true ecosystem-based management approach will require that the entire watershed be included in the protected area's sphere of influence—even if direct inclusion is not possible. How much of the watershed should be included to capture critical ecological processes is a function of the dynamics of the system.[11]

A similar consideration in a coastal protected area is how far out to sea the bounds of the protected area should extend. All marine and coastal habitats in any given ocean basin are somehow linked, but including whole oceans in protected area plans is most certainly not a realistic proposition. Nevertheless, strong linkages to pelagic systems will necessitate expansion of the protected area to include some open ocean features.[4] In addition, there may be sociopolitical reasons to extend the bounds of the protected area seaward, such as security/enforcement concerns, the need to resolve offshore fisheries disputes and others.[12]

## SECTION 3. ASSESSING THREATS TO COASTAL AND MARINE BIODIVERSITY

Given the enormous range of human activities that can impact coastal and marine ecosystems in potentially negative ways, and given that the magnitude of threat varies not only according to the intensity of the activity but also the condition of the ecosystem and the existence of other, possibly cumulative threats, is it possible to objectively assess which threats should take priority in a marine protected area management scheme? The answer is yes–but the level of analysis required and its inherent objectivity will be a function of the scale on which the site selection is occurring. For instance, an assessment of threats that helps a global environmental body or non-governmental body establish priorities for marine protected area support will take place on a much rougher scale than an assessment of threats undertaken by a local planning office, which needs to know where to site the next marine protected area. In the end, however, the criteria used to evaluate threats and establish priorities at any geographical scale will be similar.

The highest priority threats to marine biological diversity and the ecosystems that maintain it are those which cause irreversible change or damage. Habitat conversion, such as the development of wetlands areas for housing or industry and the destruction of coral reefs for the construction of ports are all impacts that are irreversible (if not absolutely so, then so costly that restoration is not a feasible option). Causing a species to go extinct or an entire population of organisms representing a particular genetic makeup to be extirpated are also irreversible impacts.

Other impacts that should be rated as high priority include those that are geographically far-reaching or exist for long periods of time. For instance, the construction of a coastal engineering project that alters the physical processes in the region is a geographically large scale impact–the same is true for large scale damming and irrigation projects (as described in chapter 2). Other high priority threats may be low-level in terms of the change caused at any point in time, but when they continue for long periods or become chronic, they should be taken seriously.

Ranking threats on a one-by-one basis is relatively easy–however, getting the big picture in terms of assessing the magnitude of cumulative threats is not. The simplest way to attempt to do this is to rank individual threats and then display them all graphically–overlaying threats to a given area to show where threats to marine biodiversity are most highly concentrated.

## SECTION 4. IDENTIFYING SPECIAL OPPORTUNITIES FOR MARINE PROTECTED AREA DESIGNATION

Global, regional and national planners have the unenviable task of developing rigorous, non-subjective ways to evaluate where and when and in what form management is needed. But even in the most forward thinking, well-developed planning process, there is a need to be able to accommodate special needs as they arise. Such needs and opportunities cannot be anticipated. Yet often the best rationale for developing and implementing a marine protected area is that people want it: local communities, their representatives, even a dedicated individual. The role of dedicated people—stewards for the oceans and for the protected area—cannot be undervalued. In fact, the most successful marine protected areas are most often those for which a single person or group became a staunch supporter and forced it to succeed.

How, then, is it possible to produce a broad-brush overview of a region in order to identify which areas may be more important than others? Ideally, scientific information on system ecology, utilizing physical and biological oceanography and hydrology, can provide answers on where critical ecological processes are concentrated when. Even rudimentary information on currents, mixing, nutrient loading and bottom topography can provide salient information on critical areas. Such areas of interest can be mapped conventionally or using Geographic Information Systems (GIS) technology in order to provide the baseline for area identification. In this way, if an opportunity to establish a marine protected area emerges, its potential importance to overall marine conservation can be readily assessed.

When scientific information concerning bathymetry, currents, and other basic features of the environment are patchy or difficult to obtain, indicator species or other surrogates can be used to determine the importance of an area. The distribution and abundance of keystone species, if they have been identified as such, often provide basic understanding of ecosystem function. In some areas where scientific research is sporadic or non-existent, data on use of commercially or culturally important species may be used as a rough approximation of where some critical areas can be found (e.g., see Papua New Guinea Conservation Needs Assessment; Agardy and Pernetta, 1992).[3] However, when using such surrogates to identify priority areas for protected area work, it is important that physical features of the environment be coupled as well as possible with information on the distribution of such "special" species. This way the species of special concern can be used as a peg for marine conservation—providing a unique and possibly unanticipated opportunity to allow the establishment of a protected area.

## REFERENCES

1. Gubbay, S. Marine Protected Areas: Principles and Techniques for Management. London, Chapman and Hall, 1995.
2. Agardy, T. The Science of Conservation in the Coastal Zone: New Insights on How to Design, Implement and Monitor Marine Protected Areas. Proc. of the World Parks Congress 8-21 Feb. 1992, Caracas, Venezuela. IUCN, Gland, Switzerland. 1995.
3. Agardy, T. Accommodating ecotourism in multiple use marine reserves. Ocean and Coastal Management 1993; 20:219-239.
4. Ray, G.C. Sustainable use of the ocean. In: Changing the Global Environment. New York, Academic Press, 1988:71-87.
5. Norse, E. Global Marine Biological Diversity. Washington, DC, Island Press, 1993.
6. United Nations Environment Programme (UNEP). Guidelines for the selection, establishment, management and notification of information on marine and coastal protected areas in the Mediterranean. Regional Activity Centre for Specially Protected Areas, Tunis, Tunisia, 1987.
7. Kelleher, G., C. Bleakley and S. Wells. A Global Representative System of Marine Protected Areas. Washington, DC, World Bank, 1995(4 volumes).
8. Agardy, T. and J. Pernetta. A preliminary assessment of biodiversity and conservation for coastal and marine ecosystems in Papua New Guinea. In: Papua New Guinea Needs Assessment. Biodiversity Support Program., US Agency for International Development., Washington, DC. Vol. 2 1992:381-421.
9. Carter, R.W. Coastal Environments. New York, Academic Press, 1988.
10. Chapin, F.S. III, E.D. Schulze and H.A. Mooney. Biodiversity and ecosystem processes. Trends in Ecology and Evolution 1992; 7(4).
11. Parsons, T.R., M. Takahashi, and B. Hargrave. Biological Oceanographic Processes. Third edition. Oxford, Pergamon Press, 1984.
12. Johannes, R.E. Traditional conservation methods and protected marine areas in Oceania. In: J. McNeely and K. Miller, eds. National Parks, Conservation and Development. Smithsonian Institution Press, Washington, DC, 1984; 344-347.

= CHAPTER 10 =

# GUIDELINES FOR MARINE PROTECTED AREA DESIGN AND MANAGEMENT PLANNING

## SECTION 1. A GENERIC PLANNING PROCESS FOR MARINE PROTECTED AREAS

In recognition of the fact that the creation of conventional marine parks has done little to abate directed overexploitation and indirect but chronic degradation of vital coastal systems, marine planners have begun to develop principles for scientifically-based multiple use areas. These multiple use areas aim to protect those parts of the comprehensive coastal/marine ecosystem that are most ecologically critical–targeting the vital organs, as it were. Two kinds of basic questions need to be answered to provide the basis for such planning: 1) where and when critical processes such as nutrient loading, feeding, spawning and breeding, and migration are concentrated; and 2) which areas (and the ecological processes that they support) are most at risk from current or prospective human use.

In coastal and marine systems, scientists have begun to identify these critical or driving processes in the physical, geochemical, and biotic realms. Important features of the ecosystem which contribute to productivity, diversity and resilience of systems include such things as upwelling, longshore and tidal fronts, warm and cold core rings, currents, freshwater input and mixing, nutrient loading and transport, atmospheric exchange, population recruitment, the existence of keystone species, symbiotic associations and predator/prey relationships.[1,2] This may seem like an impossibly wide spread of parameters for investigation, but our level of understanding is advanced enough to know that in certain systems, a few identifiable features may be the controlling factors.[3] Limiting substantial negative impacts on those critical

*Marine Protected Areas and Ocean Conservation,* by Tundi Spring Agardy.
© 1997 R.G. Landes Company.

processes lessens the chance that we will impair the homeostatic capability of the system to maintain itself.[4]

Thus it is central to efficient coastal and marine park planning to identify where ecologically critical processes are concentrated in space and time. When these areas are well protected from both direct and indirect degradation, e.g., as core areas within a multiple use zoning plan, the marine system can continue to thrive, if relatively pristine, or begin its return to health and productivity, if degraded.[5] Core and buffer zonation is thus useful in helping to protect the most vital processes while allowing optimal utilization of resources to meet human needs.

The planning of marine protected areas can be done on a case-by-case basis as opportunities or needs arise, or it can be undertaken in the context of national strategic planning. National level efforts to conserve marine biological diversity and protect marine resources for future use must encompass five steps. The first step is assessment: determining not only the status of the natural resource base and priorities for management, but also the current potential for government-based regulation and public acceptance of conservation measures. The second step is prioritizing problems, by identifying the root causes underlying marine degradation and user conflict. The third step involves developing a plan which utilizes the most effective conservation measures available. Such a plan will involve defining strategic objectives (as specifically as possible), developing ways to guarantee continued involvement of all stakeholders, setting realistic timetables, and developing ways to monitor progress towards achievement of specific objectives. The fourth step is the integration of these conservation measures across all sectors, so that management is coordinated and maximally responsive. The fifth and final step brings us back to the beginning of the process: periodic evaluation to monitor the effectiveness of management–and subsequent adaptation of management so that human needs can continue to be met as social and environmental conditions change.[6]

In all of the above steps, while the initiative should be taken by governments and planned according to national scales, participation by those depending on the resources is necessary to the eventual efficacy of management.[7] Involving stakeholders in marine conservation only late in the planning and implementation processes can incur several types of cost: 1) traditional knowledge is not available for assessments, 2) acrimony by being excluded can lead to even bigger resource use conflicts, and 3) local people may not be predisposed to shoulder a part of the burden for monitoring or managing resource use.

The following table (Table 10.1) lists the types of information that planners and decision-makers should consider when developing an action plan for marine protected areas in order to conserve marine biological diversity.

**Table 10.1. Considerations for assessing the state of knowledge about the marine environment**

* Have inventories been performed to assess species and habitat richness in coastal zone and marine areas under national jurisdiction?
* Has analysis been undertaken to identify hot-spots of high species diversity?
* Has coastal and marine diversity been mapped?
* What information exists on biological indicators of diversity and ecosystem health?
* Are geographic data available on biological productivity of the coastal and marine areas?
* What information can be derived about proximate threats to marine biological diversity?
* Do good indicators exist for pinpointing direct impacts on biodiversity (overharvesting of living marine resources, habitat alteration for coastal residential development, port development, siting of industry, etc.)?
* What measures exist for assessing the magnitude of indirect impacts on coastal and marine biodiversity (non-point source pollution such as urban run-off, agricultural run-off, riverine inputs of agricultural nutrients and toxics, changes to hydrological regimes, etc.)
* Are customary tenure or ownership of marine areas/resources practiced and, if so, are such practices acknowledged by the authorities as being valid?
* How reliant are coastal populations on living and non-living marine resources?
* Does the oral history, written historical documentation, or mythology of the indigenous culture provide information about marine resources and their past use?

Once information about local resources and their use is compiled, an accurate picture of the target ecosystem can be assembled. This will provide the background information for any science-based marine or coastal protected area to be established.

The actual planning and design of a marine protected area is very much a sociological process, not a purely scientific one. For one, the design of the reserve will be a function of the objectives it serves, and these objectives are borne of people's needs and expectations. Second, the planning of a management regime cannot take place in a vacuum—after all, it is people and their activities that we seek to manage, not ecosystems per se. Stewardship and responsibility cannot be expected to develop for a protected area that is imposed on users of marine resources without their involvement.

Every marine protected area will differ in terms of the best way to approach planning and implementation. However, a generic process can be outlined that should guarantee that user groups are fully involved and that science is harnessed appropriately. Such a generic process is outlined in Table 10.2.

**Table 10.2. A seven step process for designing a marine protected area**

1. Identify and involve all user groups
2. Set realistic goals for the protected area with all concerned parties (these become the protected area-specific objectives)
3. Design the outer bounds of the marine protected area to reflect objectives
4. Develop a preliminary zoning plan within the outer boundary
5. Amend zoning to reflect user group expectations and needs
6. Formulate the management plan
7. Monitor the marine protected area to ensure that scientific and sociological goals are being met, distribute the evaluations widely, and periodically adapt the management plan for maximum effectiveness

## SECTION 2. DEVELOPING MANAGEMENT APPROPRIATE TO NEEDS

What options exist for protecting marine areas? Coastal and marine habitats and organisms can be conserved through the establishment of any number of protected area types, including marine parks, marine extensions of terrestrial parks, marine sanctuaries, multiple use areas, and biosphere reserves. In general, the larger the protected area and the more uses that are accommodated through careful planning, the better conservation can be served.

There are many concrete incentives for creating marine protected areas, particularly in East Africa and South Asia.[8] Immediate incentives include the promise of international aid for research, planning and management.[9] Marine protected areas also provide host countries with a certain amount of international prestige, and significant marketing potential for tourism.[10] Over the long term, marine protected area designation makes possible both sustainable use and sustainable conservation.

To be truly effective, integrated management of marine protected areas should address, or at least consider, four general conservation goals. First, where pressures to exploit resources are high, coastal management should adhere to a regime that is based on scientifically sound, rational definitions of what levels of use are truly sustainable. These sustainable use levels must be identified with the entire system in mind, not a single species at a time. Second, integrated coastal management should confer special management attention to those components of the ecosystem (species or processes) that are highly threatened. Endangered species such as endemic fishes, marine mammals, sea turtles, coastal forests, etc., are useful indicators of general trends in environment condition, and require special management. Third, coastal management should, wherever possible, utilize some type of zoning system to protect habitats that act as critical areas for the ecosystem in question. Even low diversity areas with functional significance to the ecosystem should receive priority protection. Lastly, specific conservation measures, since they do not occur in a vacuum but rather exist in the context of a wider matrix of differing management regimes, must tackle the problems of indirect degradation of the target ecosystem. They should act to focus scientific and political attention on the indirect degradation of target areas through point source pollution, uncontrolled run-off and poor watershed management, inconsistent coastal use outside the managed area, and global change. Only in this way can integrated coastal management be truly comprehensive and integrative.

Marine area management in the U.S. and in some other industrial countries has emphasized the restructuring of administrative frameworks to make control of all marine/coastal resource use and space as coordinated as possible. While such administrative coordination is an important and laudable goal, coastal management through marine protected areas in many other parts of the world require more than amending extant regulatory systems. Effective conservation of coastal ecosystems worldwide requires the following: 1) deciphering and articulating people's needs and thus their potential reliance and impact on coastal systems; 2) using the best available scientific information to determine what levels and what kinds of resource use are optimal; and 3) developing a management structure that accommodates a wide range of needs with minimum user conflict and empowers local communities to be wise stewards of the seas.

The first of these three points relates to goal-setting, which should be a dynamic, ongoing process. Elaborating specific objectives for management, whether under the aegis of a national/provincial/state coastal management plan or a multiple use protected area, forces objective evaluation of people's requirements, expectations, and future impacts on coastal systems. The importance of canvassing all users in all segments of society to get at this goal-setting cannot be overstated. Stakeholders can be found in various levels of government, businesses, and local communities who may depend on coastal resources for subsistence or market economies. Other, less-directly linked stakeholders include non-governmental organizations and multilateral aid agencies or other investors. All of these stakeholders have interests—sometimes conflicting—that must be considered in the development of a realistic and equitable protected area management plan.

The second point relates to harnessing the science we currently have available to us to develop rigorous management, in turn ensuring that use of marine resources and ocean space will be sustainable over long time frames. This science should draw from advances in the fields of ecology, population dynamics/genetics, physical oceanography and hydrology and environmental studies.

The third introductory point relates to socio-political concerns. It is foolish to think we can develop a generic model for protected area management that will take root wherever we plant it. Political, cultural and economic differences around the world will mean not only that objectives will vary from place to place but also that the mechanisms by which those goals are reached will likewise be diverse. Adapting management so that it is effective requires a thorough understanding of local conditions: sociological as well as environmental. Top-down approaches aimed at developing large scale frameworks for coastal management must be complemented by bottom-up approaches that root this management in local realities.

Marine protected areas, whether in the form of small marine parks or extensive multiple use areas, provide a valuable avenue for protecting critical areas quickly. A marine protected area gives us a geographic framework in which to make resource use sustainable, provides a tangible asset to local people and governments, forces the articulation of an institutional structure in which intelligent, forward-thinking and judicial development decisions can be made, and establishes a concrete set of reference points against which we can assess the condition of the marine environment and monitor the sociological, economic, and environmental success of our efforts. Thus marine protected areas may be the most important means at our disposal to begin turning the tide and stemming problems of marine and coastal degradation.

Multiple use, multi-objective marine protected areas are but one category of small scale models demonstrating the kind of integrated marine resource management we should be practicing on regional and even global scales.[11] Marine protected areas help coordinate management activity so that it is optimally effective, and help satisfy human needs by ensuring that renewable resources are sustainably used without undue user conflict. Well-planned marine management, including that accomplished in marine protected areas or in Integrated Conservation and Development Projects, will serve local communities as well as national economies.

## SECTION 3. MONITORING AND EVALUATION

Monitoring and evaluation must strive to tell us three things: a) the current environmental condition or status of the ecosystem or its components; b) trends in condition or predictions for the short-term future of ecosystem components; and c) the efficacy of management and appropriateness of conservation measures. From a functional viewpoint, monitoring can be grouped into three different classes: 1) estimation of ecosystem condition and magnitude of impairment of ecological function; 2) measurement of habitat diversity and changes in landscape and seascape ecology; and 3) measurement of species diversity. A functional approach is critical, since ultimately conservation is concerned not with maintaining the static structure of ecosystems, but rather with maintaining their dynamic productivity and their potential for adaptation.

The distinction between basic ecological research and applied monitoring is that the latter aims to chart progress and indicate when management changes are needed. The appropriate choice of parameters, as well as the design of the monitoring plan, will ensure that research is useful. Table 10.3 lists some of the parameters that one might use to measure the conservation progress made by coastal or marine protected area management.

*Table 10.3. Monitoring Parameters and Tools to Measure Them*

**Species Diversity in Selected Zones of Interest**

| | |
|---|---|
| Plant communities | Vegetative transects |
| Ornithological | Bird counts |
| Marine biodiversity | Fish counts and marine plots |
| Introduced species | Census for distribution/abundances |

**Habitat Diversity**

| | |
|---|---|
| Forest fire damage | Aerial photography & GIS |
| Habitat alteration | Aerial photography, ground-truthing & GIS |
| Fragmentation | Mapping of roads, transects to determine edge effects |

**Impairment of Ecosystem Function**

| | |
|---|---|
| Aquifer recharge | Water level measurements, groundwater salinity readings, etc. |
| Soil mineralization | Soil quality measurements, transects to determine changes in vegetation, etc. |
| Hydrological patterns | Measurement of fluxes in drainage and flows, mapping, etc. |
| Lacustrine or riverine pollution | pH, redox potential, oxygen levels, conductance, D.O. saturation, resistivity, trace metal analysis |
| Eutrophication in nearshore waters | Secchi disk measurements, chlorophyll counts, aerial surveys of blooms and fish kills (when applicable), etc. |
| Debris accumulation, sources and sinks | Beach and wetlands surveys to quantify amount of debris, its nature, and qualitatively assess impacts, etc. |
| Beach use impacts | Census beach use, map beachfront erosion due to trampling, map coastal engineering projects, etc. |
| Overexploitation of key natural resources | Port sampling, catch per unit effort estimates, surveys of magnitude of sand/gravel extraction, etc. |
| Depressed redox potential of wetlands | Leaf litter, plant biomass, evapo-transpiration estimates |

Whatever specific parameters are chosen and whatever detailed methodology is used, the monitoring plan will have to consider the following features:

* **Controls**: Since monitoring is a rigorous applied science that follows established conventions of experimental ecology, it will be important to establish control sites wherever possible. These controls provide the basis for qualitative comparison and quantitative, statistical analysis. Examples of controls include remote, inaccessible beach areas for use in monitoring human impacts on tourist beaches, and pristine dune areas with no recent fire damage or other anthropogenic impacts for use in monitoring dune condition. It is assumed that no "true" controls, devoid of human impacts, will be identified, nonetheless the best possible controls should be utilized at the very conception of the monitoring program.

* **Redundancy**: For best results and most rigorous analysis, several monitoring stations should be set up to measure change in the same types of habitats. This redundancy will ensure that natural stochastic changes particular to an area or specific ecological community will not skew results. Of course, the number of redundant monitoring sites and the frequency of monitoring activity will be influenced by time and funding constraints–an optimal regime will be one that is scientifically sound yet cost-effective.

* **Consistency**: Consistency refers not only to methodology and tools employed in the monitoring, but also, to the extent practicable, personnel involved in data gathering and analysis. Again, the benefits presented by having consistent, reliable data are contrasted to possible logistic and other costs. One negative side-effect of maintaining continuity in personnel is that training and development of field experience will be limited to a few people.

* **Repeatability**: Repeatability is similar to methodological consistency, although targeting parameters for which measurements are easily repeated involves careful planning of methods and not training, as above. Many new developments in environmental technology de facto increase repeatability, but the high capital costs of acquiring such technology tools must be weighed against the benefits accrued by using them.

* **Feasibility**: Without considering feasibility when designing the monitoring regime, the exercise can become a waste of time and resources. Feasibility pertains not only to technical capacities and costs, but also to political and social considerations.

Some questions that monitoring and evaluation programs might target include the following:

> Are there significant changes in patterns of species distribution and abundance?
>
> Are productivity levels unaltered by use?
>
> Are core/buffer or other zones correctly placed?
>
> Are seasonal management measures effective?
>
> Are the outer bounds of the management area sufficient to conserve the target area as an ecosystem?

Have user conflicts been avoided/reduced?

Do users understand and respect the zoning plan?

Do communities perceive a benefit from the marine protected area designation?

Is the protected area economically sustainable or profitable?

Are local users politically more empowered and thus more inclined to become stewards of the resource rather than just harvesters?

* **Information exchange:** linking to a network of scientifically-based marine protected areas

## SECTION 4. TEN PRINCIPLES FOR MARINE PROTECTED AREA SUCCESS

Every marine or coastal protected area is unique–a special combination of environmental characteristics, societal needs, and management frameworks. As repeatedly stressed throughout this book, no model exists that is appropriate in all circumstances. However, there are important general principles for living resource conservation that can guide any marine protected area process. These general principles (Mangel et al. 1995) are appended (Appendix).[12] More specific guiding principles, specific to marine protected area design and management, are listed below:

1. **Clearly define specific objectives for the marine protected area at the onset,** with as much input from all stakeholders as is possible.

2. **Design zoning to maximize protection for ecologically critical areas,** while allowing sustainable use in less sensitive, vulnerable, or important areas.

3. **Design marine protected area boundaries so that they reflect ecological reality** (avoid squares and other "unnatural" shapes, encompass estuaries and landward sides of coastal zones, etc.)—and be prepared to alter the design or the management as more ecological and sociological information becomes available.

4. **Design the marine protected area and develop its management plan with feasibility in mind**—and look for ways to self-finance management operation from the onset.

5. **Make the planning process truly participatory,** as opposed to allowing user groups to comment on a plan developed by a single stakeholder (usually a government agency).

6. **Use the marine protected area as a way to raise awareness,** educate, and empower—not just at the beginning of

the planning process but throughout the protected area's lifetime.

7. **Develop monitoring and evaluation methodologies that are appropriate to the specific objectives** and include these functions in design criteria.

8. **Form an independent, non-partisan or multi-user group body to manage the marine protected area** and monitor its effectiveness using established benchmarks.

9. **Undertake valuation exercises periodically** to ensure that the full value of the marine protected area is being realized in order to provide incentives for the establishment of additional marine protected areas.

10. **Use individual marine protected areas as a starting point for more effective marine policies overall**—either to begin a representative network of MPAs on a national scale, or to draw attention to larger scale environmental problems such as land-based sources of pollution, regional overexploitation or habitat destruction, etc.

## REFERENCES

1. Caddy, J.F. and G.D. Sharp. An Ecological Framework for Marine Fishery Investigations. FAO Fish. Tech. Pap. 1983; 283.
2. Gulland, J.A. and S. Garcia. Observed patterns in multispecies fisheries. In: R. May, ed. Exploitation of Marine Communities: Report of the Dahlem Workshop. 1-6 April 1984, Berlin. Springer Verlag Life Sciences Research Report 1984; 32:155-190.
3. Moore, J.C., P.C. de Ruiter and H.W. Hunt. Influence of productivity on the stability of real and model ecosystems. Science 1993; 261:906-907.
4. Barrett, C.W., G.M. Van Dyne and E.P. Odum. Stress ecology. Bioscience 1976; 26(3):192-194.
5. Kriwoken, L.K. The Great Barrier Reef Marine Park: an assessment of zoning methodology for Australian marine and estuarine protected areas. Maritime Studies 1987; 36:12-21.
6. Holling, C.S. Adaptive Environmental Assessment and Management. New York, John Wiley and Sons, 1988.
7. World Bank. World Bank Participation Sourcebook. Environment Department, Social Policy and Resettlement Policy Division, Participation Series Working Paper 1996; 19.
8. Pernetta, J. Marine Protected Area Needs in the South Asian Seas Region: Vols. 1-5. IUCN Marine Conservation and Development Reports, Gland, Switzerland, 1993.
9. Organization for Economic Cooperation and Development (OECD). Guidelines for Aid Agencies on Global and Regional Aspects of the Development and Protection of the Marine and Coastal Environment. Guidelines on Aid and Development 1996 No. 8. Paris, France.

10. Pearsall, S. *In absentia* benefits of nature preserves: A review. Envir. Cons. 1984; 11:3-10.

11. Burbridge, P.R., N. Dankers and J.R. Clark. Multiple use assessment for coastal management. Coastal Zone 1989;89:33-45.

12. Mangel, M., L.M. Talbot, G.K. Meffe, M.T. Agardy, D.L. Alverson, J. Barlow, D.B. Botkin, G. Budowski, T. Clark, J. Cooke, R.H. Crozier, P.K. Dayton, D.L. Elder, C.W. Fowler, S. Funtowicz, J. Giske, R.J. Hofman, S.J. Holt, S.R. Kellert, L.A. Kimball, D. Ludwig, K. Magnusson, B.S. Malayang, C. Mann, E.A. Norse, S.P. Northridge, W.F. Perrin, C. Perrings, R.M. Peterman, G.B. Rabb, H.A. Regier, J.E. Reynolds III, K. Sherman, M.P. Sissenwine, T.D. Smith, A. Starfield, R.J. Taylor, M.F. Tillman, C. Toft, J.R. Twiss, Jr.,J. Wilen, and T.P. Young. Principles for the conservation of wild living resources. Ecol. Appl. 1996; 6(2):338-362.

# PART V:
# FUTURE PROSPECTS

PART V:
FUTURE PROSPECTS

# INTERNATIONAL INSTRUMENTS AND FRAMEWORKS FACILITATING MARINE PROTECTED AREA DESIGNATION

## SECTION 1. INTRODUCTION TO FRAMEWORKS AND LEGAL INSTRUMENTS

The world's coastal zones are in trouble: we have befouled our precious shores and the waters beyond. The earth's coastal belt, extending from the inland margin of the coastal plain to continental shelf waters offshore, accounts for only 10% of the earth's surface–yet it supplies more than 30% of the world ocean's immense productivity and accommodates over two-thirds of the human population with living space. Although coastal areas provide critical food, transportation, recreation, and energy resources to increasing numbers of people each year, the resource base itself and the ecosystem which maintains it are now overburdened and jeopardized by both careless use and extrinsic impacts. Recent widespread algal blooms, fish kills and the decimation of coastal species and habitats are testimony that this vital ecosystem is now heavily stressed worldwide.

What is happening to our coasts and nearshore seas is the result of conflict: the conflicting needs and desires of local inhabitants, industries, and governments. At the root of this conflict is a diversity of opinion on who has rights to what, and a disparity of views about whether the environment should be exploited today or preserved for tomorrow. In many, if not most, parts of the world, this coastal conflict draws battlelines between three groups of people: 1) local inhabitants who wish to maintain traditional uses of marine and coastal

*Marine Protected Areas and Ocean Conservation*, by Tundi Spring Agardy.
© 1997 R.G. Landes Company.

resources; 2) tourists and the industry which accommodates visitors, which seek to ever-increase access to coastal areas and resources; and 3) industrialization, which threatens to transform coastal landscapes into urbanized areas. The voice of the first group is weakest, and easily overpowered by the other, more environmentally harmful, groups.

In much of the world, environmental destruction is most chronic and acute in the coastal margins. The degradation of coastal environments is not merely an aesthetic consideration, but instead interferes with our ability to utilize and enjoy coastal resources. The negative impacts are sometimes direct, as in the case of overfishing, ocean dumping, coral reef blasting and other forms of habitat alteration or destruction. They may also be indirect, as in the case of ecosystem poisoning via terrigenous run-off and river-borne pollution, and eutrophication caused by nutrient overloading. Unfortunately, the vastness of the ocean and its apparent bounty obscures the fact that both renewable and non-renewable resources are indeed limited, and now threatened.

Despite the obvious importance of coastal systems, our track record in managing and mitigating our impacts on them is not laudable. In fact, coastal zone management may sport the poorest record of all conservation efforts worldwide. The reluctance, or perhaps more fairly, the inability to defend marine resources and traditional uses of these resources with the same vigor as is directed towards terrestrial ecosystems has to do in part with the fact that marine resources are regarded as a global commons. Lack of private ownership or clear sovereignty complicates attempts at delimiting management areas. Administratively, this means that issues of sovereignty and the right to control use and manage exploitation is sometimes unclear. In any given coastal area, local authorities may be in conflict with provincial and/or national governmental bodies in how to manage resource use.

We also have less contact with marine resources and the effects that human activities have on them—and what is out of sight is often out of mind as well. Although a great whale may be a cherished symbol of our reverence for the seas, for instance, a sick or dying whale escapes notice and rarely enters the consciousness of the public. This means that there is relatively less public concern for conserving marine areas, in comparison with the much-publicized and widely embraced effort to protect tropical forests or endangered terrestrial species.

This already difficult situation is compounded by the paucity of knowledge about marine ecosystem dynamics, making ecologically-based management measures difficult to identify. Although the same ecological principles may apply to terrestrial and marine systems, our understanding is complicated by differences in scale and dynamics. These differences in scale are such that marine and coastal ecosystems tend to be much larger than terrestrial systems, with widespread and complex connectivity between ecosystem components (Steele, 1974). This connectivity extends to both sides of the land-sea margin, so that offshore systems may be linked to upland systems via watershed transport and the movement of organisms. Marine and coastal systems are more dynamic than terrestrial systems, varying temporally and spatially by having few linkages to fixed substrates. These differences in scale and dynamics mean that conservation of coastal systems must target larger areas, including all possible linkages, and should be tailored to take into account complicated dynamics.[2]

As mentioned previously, part of our inability to control our use of coastal and marine resources to levels which are sustainable over the long-term has to do with the important perception that the oceans are a global commons. This is an attitude, as well as an administrative, problem. Although the sense of stewardship in some (mostly lesser developed) coastal communities is strong and has been so for many generations, competition for coastal space undermines people's personal investment in maintaining healthy coastal environments. With the vast majority of the world's population living in the coastal belt of the continents, and with the emergence of a collective perception that coastal resources can and should be taken free of social cost, anarchic overexploitation and misuse has occurred. An exploitative and competitive attitude among many users of the resource, coupled with the insidious and chronic degradation of the ecosystem as a whole, bodes ill for the conservation of many coastal areas. Widespread global change, and such prospective threats such as sea-level rise, further imperil our coasts.

We must now look to new ways to control our use of and impact on coastal areas. The management of marine and coastally-based industries on a singular case-by-case basis, and the designation of small marine parks which ignore ecological considerations, contribute little to a systematic means of reducing environmental degradation of the coasts. What is needed instead is a new perspective, one which accepts both the complexity and interconnectedness of all elements of the coastal ecosystem <u>and</u> humans as rightful users of those resources. Multiple use marine protected areas, and national systems of regional networks of such marine protected areas, will help us move in this direction. We cannot pretend that coastal population growth and the need for coastal and marine resources will decrease in the future: we can only plan that this growth will not compromise our use of the coasts and seas, nor that of future generations.

A growing number of important international conventions and other legal instruments aimed at conserving marine and coastal environments suggest the world is ever more ready to fully embrace such a philosophy. Many of these explicitly call for the formation of marine protected areas; others provide a framework for protected area establishment. The most prominent of these agreements are described in detail below.

## SECTION 2. THE UNITED NATIONS CONVENTION ON LAW OF THE SEA (UNCLOS)

The United Nations Convention on Law of the Sea (UNCLOS) recently came into force after two decades of debate and discussion among coastal nations. Ratification by over 80 nations in November 1995 brought entry into force, so that a global international treaty now codifies much of what is customary law regarding marine jurisdictions, rights to resource use, and conduct while in marine areas.[3] The UNCLOS makes only limited reference to marine area protection, but it does reinforce designations pursuant to other international agreements and expands options for protected area establishment.[4] Article 211.6 of the treaty specifically pertains to protected areas, citing the need for designation of "special areas" potentially threatened by vessel-source pollution. In addition, the International Maritime Organization (IMO) has related recommendations under its 1991 Guidelines for the Designation of Special Areas and the Identification of particularly Sensitive Sea Areas. Another related treaty, the International Convention for the Prevention of Pollution from Ships and its 1978 Protocol, also makes reference to the need for identifying and protecting special areas from marine pollution.

## SECTION 3. UNITED NATIONS ENVIRONMENT PROGRAMME (UNEP) REGIONAL SEAS CONVENTIONS

These international agreements signal multilateral efforts to manage coastal and ocean areas cooperatively in a regional framework. The Barcelona Convention covering the Mediterranean Sea and its nations is a landmark treaty which has proved that regional cooperation is possible. Under the Barcelona Convention, a protocol on specially protected areas was developed to encourage parties to the convention to establish marine protected area networks.[5] The Cartagena Treaty which applies to the Caribbean insular and mainland countries is modelled on the Mediterranean Regional Seas Programme and also lends credence to the idea that regional management of ocean space is a realistic and attractive alternative to unilateral management. Marine protected areas, including coastal biosphere reserves, have been designated thanks, in part, to the system established by these treaties.[6]

## SECTION 4. CONVENTION ON BIOLOGICAL DIVERSITY

The Convention on Biological Diversity targets biodiversity at all levels, both terrestrial and marine. The Second Conference of Parties (COP II), however, concentrated on marine and coastal biodiversity specifically. The outcome of COP II includes a non-binding document called the Jakarta mandate that stresses the importance of the Biodiversity Convention as a legal tool for promoting the conservation of marine and coastal biodiversity.[7] Under the Convention each party is obligated to act within its national jurisdiction to protect coastal and marine biodiversity. Article 8 (sections a, b and e) specifically calls for the establishment of protected areas for the conservation and sustainable use of threatened species, habitats, living marine resources and ecological processes. In recent meetings relating to the Convention (e.g., the meeting of the Subsidiary Body on Scientific, Technical, and Technological Advice), the need for national and regional networks of marine protected areas also surfaced.

## SECTION 5. UNESCO BIOSPHERE RESERVE PROGRAMME

It was in the first major period of environmental awareness, during the seventies decade, that the biosphere reserve concept was formed and fostered. The United Nations Educational, Scientific and Cultural Organization (UNESCO) launched its Man and Biosphere (MAB) Programme in 1971, and in 1974 began the Biosphere Reserve Programme as one aspect of the wider MAB philosophy.

The MAB Programme is a nationally based, international programme of research, education and information transfer which seeks to provide the scientific basis for resolving conservation and development conflicts. It emphasizes multi-disciplinary research on the interaction between social and ecological systems and applies a holistic approach to understanding the relationships between man and the environment.[8] Biosphere reserves belong to a unique category of specially-protected area that has evolved under and with the MAB Programme. Through MAB-designated biosphere reserves countries can reconcile important environmental concerns with necessary and inevitable growth and development.

The UNESCO Biosphere Reserve Programme provides a useful model to incorporate human needs into long-term planning for conservation. Central to the model is multiple use zoning to protect sensitive habitats and critical ecological processes in core areas, while allowing managed use in buffer zones. This model has particular potential in coastal areas, where conventional "garrison reserve" measures to preserve nature or protect the environment are not compatible with the open, multi-jurisdictional, and common property nature of marine systems. The successful application of the biosphere reserve model in coastal areas will require a functional perspective that recognizes all the important linkages between and within marine and terrestrial areas. A functional approach allows delineation of the outer boundaries of the protected area (making the managed area a functionally viable entity), as well as helping to highlight where critical processes that drive the system are concentrated. If such "vital organs" of a system can be protected, humans will be able to continue to reap its resources and derive benefits from its use, leading to greater economic and sociological sustainability.

Related to the Man and Biosphere Programme of UNESCO is the Convention Concerning the Protection of the World Cultural and Natural Heritage (World Heritage Convention). This convention, in force since 1972, creates international support for the protection of sites of global significance in terms of cultural or natural heritage. Obligations by states ratifying the treaty include identifying, protecting and conserving for future generations unique cultural and natural areas. A multilateral financing mechanism, called the World Heritage Fund, was established by the Convention to enable such conservation to take place in developing nations. Although most World Heritage Sites existing in the world today are terrestrial, exceptional marine protected area sites within territorial waters may and should be designated.

## SECTION 6. AGREEMENTS RESULTING FROM AGENDA 21 OF THE RIO EARTH SUMMIT MEETING

The 1992 United Nations Conference on Environment and Development (known as the Earth Summit) was held in Rio de Janeiro, Brazil. The participants at the Earth Summit adopted Agenda 21 as a program of action for sustainable development. While this instrument is not binding, parties that sign the text have a strong moral obligation to ensure its full implementation.[7] Spin-off agreements from Agenda 21 include the United Nations Environment Programme's Conference on the Protection of the Marine Environment from Land-Based Activities held in Washington, DC in 1995 and the United Nations Conference on the Sustainable Development of Small Island Developing States held in Barbados in 1994. Although neither conference stipulates the establishment of marine protected areas per se, implicit in the negotiations is the recognition that especially important or vulnerable marine and coastal ecosystems need special protection.

## SECTION 7. RAMSAR (CONVENTION ON WETLANDS OF INTERNATIONAL IMPORTANCE)

The Convention on Wetlands of International Importance (known as Ramsar since it was adopted in Ramsar, Iran, in 1971) is the first of the global nature conservation treaties and is still the only one devoted to a specific habitat type. The Ramsar Convention has been in force since 1975. Although the Ramsar Convention covers both freshwater and coastal wetlands–especially those important to migratory waterfowl, the Convention has great potential to catalyze coastal protected area development in signatory countries. Parties to the Ramsar Convention are obliged to designate at least one wetland area of international significance and take measures to protect it, and are encouraged to promote the "wise use" of all other wetlands areas within their jurisdiction. As of November 1995 there were 90 contracting parties to the Ramsar Convention, with 765 Wetlands of International Importance sites totalling 52 million hectares, many of which are coastal or estuarine in nature.

## SECTION 8. REGIONAL FISHERIES AGREEMENTS

Regional fisheries agreements such as the International Commission on the Conservation of Atlantic Tunas (ICCAT), the South Pacific Fisheries Forum and others, have not to date called for the establishment of marine protected areas. However, as harvest refugia and closed areas gain acceptance as important fisheries conservation tools, these small scale protected areas may be called for within the frameworks of regional fisheries agreements. Most regional fisheries agreements do not spell out specific points of law concerning marine protected areas, but there are occasional provisions concerning habitat protection, and the direction of these agreements is towards the kind of regional or multilateral cooperation that should enable the establishment of protected areas and networks.

### REFERENCES

1. Steele, J.H. The Structure of Marine Ecosystems. Cambridge, MA, Harvard University Press, 1974.
2. Doumenge, F. Human interactions in coastal and marine areas: present day conflicts in coastal resource use. Proc. on the Workshop of the Biosphere Reserve Concept to Coastal Areas, 14-20 August 1989, San Francisco, CA.
3. Johnston, D.M. and E. Gold. Extended jurisdiction. The impact of UNCLOS III on coastal state practice. In: T.A. Clingan, ed. Law of the Sea: State Practice in Zones of Special Jurisdiction. Honolulu, HI, Law of the Sea Inst. U. Hawaii, 1982.
4. Kimball, L. The Law of the Sea: Priorities and Responsibilities in Implementing the Convention. IUCN Marine Conservation and Development Report, Gland, Switzerland, 1995.
5. United Nations Environment Programme. Protocol Concerning Mediterranean Specially Protected Areas. Coordinating Unit for the Mediterranean Action Plan, Tunis, Tunisia, 1986.
6. United Nations Environment Programme (UNEP). Guidelines for the selection, establishment, management and notification of information on marine and coastal protected areas in the Mediterranean. Regional Activity Centre for Specially Protected Areas, Tunis, Tunisia, 1987.
7. de Fontaubert, C., D. Downes and T.S. Agardy. Protecting Marine and Coastal Biodiversity and Living Resources under the Convention on Biological Diversity. Washington, DC, Center for International Environmental Law, World Wildlife Fund and IUCN Publication, Washington, DC, 1996.
8. United Nations Educational Scientific and Cultural Organization (UNESCO). Biosphere Reserves: The Seville strategy and statutory framework for the world network. UNESCO, Paris, France, 1996.

# CHAPTER 12

# CONCLUSIONS

## SECTION 1. CURRENT EXTENT OF MPAs WORLDWIDE

Marine protected areas that have the support of local communities provide a necessary first step towards the attitude shift that is required if we are to avoid making a mess of the oceans. Education and awareness-raising are essential components of marine protected areas, as is scientific and management information exchange between areas. Most importantly, marine protected areas provide a mechanism for giving local people a sense of stewardship and control over their own futures; this in turn can only act to foster responsible attitudes towards the seas and coasts.

The explicit link between biodiversity (the whole suite of organisms in a system, rather than the additive sum of its parts) and function must be a major focus for future interdisciplinary research. Experts have been cautious about making generalizations on the value of biodiversity, since the maximization of biodiversity has not been shown to mean maximization of function or ecological services. Furthermore, the estimation of intrinsic value is necessarily value-laden and thus hard to derive.

Process-oriented conservation is one of many important tools available for our utilization. Some allocation of effort should go to targeting critical processes for conservation because we are constrained by both limited resources and limited time. As a physician targets the vital organs of his patient to maximize recovery, so this process-oriented conservation targets the ecological processes most vital to marine ecosystems. This occurs on two levels: at the broad policy level, top-down approaches should help develop the science and policy of conservation to make it more effective and rational, much the same way that the medical profession sets broad policies and research agendas for improving collective human health. On the more localized, field-based level, bottom-up approaches should aim to put conservation measures in place that protect the most vital processes, much the same way individual doctors promote the sustained health of individual patients

*Marine Protected Areas and Ocean Conservation,* by Tundi Spring Agardy.
© 1997 R.G. Landes Company.

by concentrating on the vital organs first and foremost. These two spheres of activity are not isolated, and the field projects play a critical role not only in dealing with conservation emergencies but also in demonstrating to policy-makers and funding institutions precisely how conservation derives benefits: advantages that can be couched in environmental, economic, social or ethical terms. The last two decades have witnessed an enormous proliferation of marine protected areas around the globe. These protected areas exist on a spectrum ranging from the very small and specialist marine parks that are established with a single objective in mind, to very large multiple use areas with complex zoning plans and multiple objectives. The most recent (1995) estimates of marine protected areas show that the bulk of existing marine protected areas occur in the Australia/New Zealand region, the Northwestern Pacific, the Northeastern Pacific and the wider Caribbean.[1] These figures, however, do not take into account those coastal protected areas that do not have sub-tidal components, which would greatly increase representation in many regions.

The 1995 Kelleher et al. report cites some 1,306 existing marine (subtidal) protected areas. The geographic distribution of these protected areas is given in Table 12.1 below:

**Table 12.1. Existing marine protected areas by region**

| | |
|---|---:|
| Antarctic | 17 |
| Arctic | 16 |
| Mediterranean | 53 |
| Northwest Atlantic | 89 |
| Northeast Atlantic | 41 |
| Baltic | 43 |
| Wider Caribbean | 104 |
| West Africa | 42 |
| South Atlantic | 19 |
| Central Indian Ocean | 15 |
| Arabian Seas | 19 |
| East Africa | 54 |
| East Asian Seas | 92 |
| South Pacific | 66 |
| Northeast Pacific | 168 |
| Northwest Pacific | 190 |
| Southeast Pacific | 18 |
| Australia/New Zealand | 260 |
| Total | 1,306 |

The total coverage for the nearly 1,000 (991) marine protected areas around the world for which areal extent is known, the total coverage is on the order of some 80,000,000 hectares. Surprisingly, three marine protected areas (the Great Barrier Reef Marine Park in Australia, the Galapagos Marine Park and the Netherlands' North Sea Reserve) account for nearly half of the existing marine protected area coverage by area.

The large number of marine protected areas in the world may be misleading, however. Many of these subtidal protected areas are really paper parks, with no management plan and very little on-the-ground (or in-the-water) management.[2] The authors did appraise management efficacy on a gross level and found that of the 1m306 existing marine protected areas, only 117 had a management level that they label as "high"—i.e., managed with significant effort and resources. It cannot be determined how many of the several hundred proposed marine protected areas identified by the authors around the globe will become paper parks as opposed to operationally effective protected areas.

## SECTION 2. LESSONS LEARNED FROM EXISTING MARINE PROTECTED AREAS

To date, few coastal nations have systems in place to establish networks of marine protected areas for achieving comprehensive marine conservation. Noteworthy exceptions to this include the Philippines, Canada and Australia. The latter two countries have planned to establish comprehensive networks for representative marine protected areas and are now in the process of implementing them. These networks are based on the premise that the best way to ensure that marine and coastal biological diversity is conserved is to establish systems of protected areas that represent every major type of coastal or marine habitat.

Interestingly, there seems to be no relation between the strength of a national plan for the development of marine protected areas and the success and efficacy of any individual marine protected area. Instead, if anything, an inverse correlation seems to exist such that the best examples of marine protected areas occur in countries with little strategic planning for national systems of reserves.[2] Rather than administrative commitments to marine protected areas and strong capacities for managing marine areas, the single most important factor underlying whether or not a marine protected area will be successful and beneficial to a country and its inhabitants is the presence of a dedicated individual or group of individuals to carry it forward (Fig. 12.1).

There are two main driving forces for initiating marine and coastal protected areas according to a systems plan—opportunities and threats. From a positive viewpoint, marine conservation projects can and should be supported in biologically rich or diverse areas and in areas where their demonstration value can be maximized. On the other hand, conservation is most urgently needed where poorly planned development, overuse of resources, or indirect degradation threaten to undermine the very resource base on which coastal peoples depend. Evaluating threats and opportunities in a systematic way can provide a means to

*Fig. 12.1. Stewardship takes many forms: scuba diver with black fan coral.*

direct marine conservation support so it is maximally effective and long-lasting.

## SECTION 3. RELATION OF MARINE PROTECTED AREAS TO BROADER COASTAL MANAGEMENT EFFORTS

The development and implementation of effective measures to manage marine protected areas is at least as important as the selection of sites and design of individual protected areas. The worldwide proliferation of marine parks that appear to be worthwhile attempts to tackle marine conservation problems yet are nothing more than paper parks suggests that this is not a universally accepted premise. Too often, agencies with jurisdiction over coastal and marine areas will devote significant resources to protected area planning and insufficient subsequent resources to development of a carefully constructed, detailed management plan or to implementation measures such as enforcement, boundary delimitation, and the like. As a result, of the 640 existing and proposed marine protected areas identified worldwide in 1995 (Kelleher, et al, 1995), only a few dozen can be considered well-managed, effective conservation regimes.[1]

Coastal conservation cannot take place in a vacuum; pre-existing claims, conflicts, and governmental jurisdictions are all part of the socio-political context which must be taken into account. Many countries have already begun the institution-building process necessary to make a comprehensive coastal zone program a reality, and in some places national coastal zone management plans exist. Whatever the state of existing coastal management, it is imperative that coastal conservation efforts, including biosphere reserve planning and nomination, work with existing agencies and legislation, rather than counter to it.

Most coastal nations adhere to a jurisdictional regime that differentiates inland seas where jurisdiction is complete, from territorial seas, and these from fishery conservation zones and EEZs where jurisdiction is limited. Management of resources in inland areas are the sole responsibility of the nation and/or state or province in which the resources are located. Territorial seas, which commonly extend from 3 to 12 miles from the coastal baseline, are areas controlled by the nation but where some freedom of navigation and other rights are allowed foreigners. In EEZs and Fishery Conservation Zones, which commonly extend 200 miles seaward of the baseline, nations only exercise jurisdiction over certain, explicitly stated, activities.[3]

Countries with established coastal zone management plans attempt to practice integrated coastal conservation through fisheries regulations, laws concerning polluting, and through the establishment of managed fishing areas, marine sanctuaries and parks, and coastal recreation areas. Although the general philosophy of these plans is similar to that of marine protected area management plans, coastal zone management plans often fall short of meeting their objectives because they are superimposed on existing patterns of use without their recognition. Resource use conflicts are rarely anticipated, and local users rarely have a voice in the planning or management process. As national coastal zone management programs evolve and become more functional, entire territorial sea areas will begin to be managed as vast marine protected areas or coastal biosphere reserves. In the meantime, smaller scale protected areas will have to provide countries with working examples of how this can be accomplished.

One reason that coastal zone conservation is light-years behind terrestrial environmental protection is that few proven models for comprehensive management under coastal or marine protected areas exist to date. And this is, in turn, true because marine resource managers have too often looked to terrestrial models for application in coastal areas. Coastal biosphere reserves and other marine protected areas provide a better tool for conservation in demonstrating not what a specific management regime should look like but instead, how to go about the conservation process.

The development of effective marine conservation and integrated protected areas planning has not been steady and gradual, but rather, like the evolution of life itself, punctuated by radical new ideas. We have long left the period of stasis behind, and are now fully in an era of dynamic and exciting new direction.

## REFERENCES

1. Kelleher, G., C. Bleakley and S. Wells. A Global Representative System of Marine Protected Areas. Washington, DC, World Bank, 1995 (4 volumes).
2. Ticco, P.C. An analysis of the use of marine protected areas to preserve and enhance marine biological diversity—A case study approach. Ph.D. Dissertation, U. Delaware, Newark, DE, 1995.
3. Kimball, L. The Law of the Sea: Priorities and Responsibilities in Implementing the Convention. IUCN Marine Conservation and Development Report, Gland, Switzerland, 1995.

# PRINCIPLES FOR THE CONSERVATION OF LIVING RESOURCES

(excerpted from Mangel et al, 1996)

**1. Maintenance of healthy populations of living resources in perpetuity is inconsistent with growing human consumption and demand for those resources.**

* Recognize that the total impact of the human presence on the living resource base is the product of human population size, per capita consumption, the impact on the resource of the extraction technologies applied, and incidental taking and habitat degradation caused by other human activities, and act accordingly.

* Recognize that if urban areas and other intense land use areas were more efficient, safer, and more pleasant, there would be a greater chance of conserving wild living resources.

**2. The goal of conservation should be to maintain present and future options by maintaining biological diversity at genetic, species, population and ecosystem levels and as a general rule neither the resource nor other components of the ecosystem should be perturbed beyond natural boundaries of variation.**

* Manage total impact on ecosystems and work to preserve essential features of the ecosystem.

* Identify areas, species, processes that are particularly important to the maintenance of an ecological system, and make special efforts to protect them.

* Manage in ways that do not further fragment natural areas.

* Maintain or mimic patterns of natural processes, including disturbances, at scales appropriate to natural system.

* Avoid disruption of food webs, especially removal of top or basal species.

* Recognize that biological processes are often non-linear, are subject to critical thresholds and synergisms, and that these must be identified, understood and incorporated into management programs.

**3. Assessment of the possible ecological and sociological effects of resource use should precede both proposed use and proposed restriction of ongoing use of a resource.**

* Identify uncertainties and assumptions regarding natural history, size, and productivity of the resource, and its role in the ecosystem.

* Identify major ecological and socio-economic uncertainties and assumptions.

* Analyze how the resource and other ecosystem components might be affected by proposed use if the assumptions are not valid.

* When available information is insufficient to make informed judgments, authorize activities contingent upon development and approval of an information acquisition plan that will ensure that the level of resource use does not increase faster than the knowledge of the size and productivity of the resource and its relation with other ecosystem components.

* Require that those most likely to benefit directly from the use of a wild living resource to pay the costs of (a) acquiring information; (b) developing and implementing an information acquisition plan; and (c) managing use of the resource.

**4. Regulation of the use of living resources must be based on an understanding of the structure and dynamics of the ecosystem of which the resource is a part and take into account the ecological and sociological influences that directly and indirectly affect resource use.**

* Allocate the use of living resources on the basis of the ecological capabilities of the species involved and their assessed values to society.

* Provide incentives to the users of living resources that correspond to the value of those resources and ensure that these incentives promote conservation.

* Ensure that institutions and property rights are consistent with conservation, including questions of tenure and access.

* Protect the welfare of future generations by ensuring that the value of biotic and abiotic resources does not fall over time.

* Recognize the possible consequences of uncertainty.

**5. The full range of knowledge and skills from the natural and social sciences must be brought to bear on conservation problems.**

 * Invoke the full range of relevant disciplines at the earliest stage possible.

 * Recognize that science is only one part of living resource conservation and is limited to investigating and objectively describing certain kinds of phenomena and processes.

 * Be prepared for unexpected events, as the natural world is highly complex and human understanding of it always contains uncertainties.

 * Require comprehensive consultations because virtually all conservation issues have economic, biological and social implications; ignoring any of these may lead to conflicts that will impair effective conservation.

 * Promote adaptive management.

**6. Effective conservation requires understanding and taking account of the motives, interests, and values of all users and stakeholders, but not by simply averaging their positions.**

 * Whenever possible, create incentives by delegating property rights to the "lowest" relevant community or societal level consistent with the scale of resource involved.

 * Develop conflict resolution mechanisms to minimize strife over resources among competing stakeholders.

 * Ally science with policy making independently of the interests of resource users.

 * Require that policy makers be held accountable for the use of the best possible data and analyses in setting policy.

 * Insofar as possible, establish agreed-upon criteria and procedures to guide decision-making on conservation measures at all levels, in order to reduce the scope of influence by political or "special" interests.

 * Ensure that formal institutions responsible for giving expression to policies and implementing conservation programs have temporal and spatial perspectives consistent with the ecological character of the resource and organizational structures that are problem-oriented, accountable, team-oriented and capable of adapting to changing circumstances.

**7. Effective conservation requires communication that is interactive, reciprocal and continuous.**

 * Ensure that communication is targeted to the audience and is based on mutual respect and sound information.

 * Require internal and external review to verify objectivity and results.

* Inform the public and motivate them to make choices about conservation.

* Develop institutions and procedures to facilitate trans-disciplinary analysis and communication that informs decision-makers.

# LITERATURE CITED

Agardy, T. Prospective climate change impacts on cetaceans and implications for the conservation of whales and dolphins. Proceedings of the International Whaling Commission Scientific Committee Symposium on Climate Change and Cetaceans, 25-30 May 1996, Kahuku, Hawaii. 1996

Agardy, T. The Science of Conservation in the Coastal Zone: New Insights on How to Design, Implement and Monitor Marine Protected Areas. Proc. of the World Parks Congress 8-21 Feb. 1992, Caracas, Venezuela. IUCN, Gland, Switzerland. 1995

Agardy, T. Advances in marine conservation: the role of protected areas. Trends in Ecology and Evolution 1994; 9(7):2676-270.

Agardy, T. Coral reefs and mangrove systems as bio-indicators of large scale phenomena: a perspective on climate change. In: Proc. of the Symp. on Biol. Indicators of Global Change, 7-9 May 1992 Brussels, Belgium. J.J. Symoens, P. Devos, J. Rammeloo and C. Verstraeten, eds. Royal Academy of Overseas Sciences, 1994: 93-105.

Agardy, T. Closed areas: a tool to complement other forms of fisheries management. In: K. Gimbel, ed. A Guidebook to Managing Fisheries Through Limited Access. Center for Marine Conservation and World Wildlife Fund, Washington, DC. 1993: 197-204.

Agardy, T. Accommodating ecotourism in multiple use marine reserves. Ocean and Coastal Management 1993; 20:219-239.

Agardy, T. Guidelines for Coastal Biosphere Reserves. In: S. Humphrey, ed. Proc. of the Workshop on the Application of the Biosphere Reserve Concept to Coastal Areas, 14-20 August, 1989, San Francisco, CA, USA., IUCN Marine Conservation and Development Report 1992, Gland, Switzerland.

Agardy, T. Last voyage of the ancient mariner? BBC Wildlife Dec. 1992: 30-37.

Agardy, T. and J.M. Broadus. Coastal and marine biosphere reserve nominations in the Acadian Boreal region: results of a cooperative effort between the U.S. and Canada. Proceedings of the Symposium on Biosphere Reserves, Fourth World Wilderness Congress, 14-17 Sept 1987, Estes Park, CO. U.S. Dept. of Interior, National Park Service, Atlanta, GA, USA. 1987

Agardy, T. and J. Pernetta. A preliminary assessment of biodiversity and conservation for coastal and marine ecosystems in Papua New Guinea. In: Papua New Guinea Needs Assessment. Biodiversity Support Program., US Agency for International Development, Washington, DC. Vol. 2 1992: 381-421.

Agee, J.K. and D.R. Johnson. Ecosystem Management for Parks and Wilderness. Seattle, WA, U.Washington Press, 1988.

Alcala, A.C. Effects of marine reserves on coral fish abundances and yields of Philippine coral reefs. Ambio 1988; 17:194-199.

Alcala, A.C. and G.R. Russ. A direct test of the effects of protective management on abundance and yield of tropical marine resources. Journal du Conseil 1990; 47(1):40-47.

Alexander, L.M. Large marine ecosystems: A new focus for marine resource management. Mar. Policy, May 1993: 186-198.

Asava, W. Local fishing communities and marine protected areas in Kenya. Parks 1994: 4(1):26-34.

Ayers, J. Population dynamics of the marine clam, *Myaarenaria*. Limnology and Oceanography 1956;1:26-34.

Bacle, J. and R. Cecil. Artisanal fisheries in Africa: Survey and research. CIDA. Hull, Canada, 1989.

Bage, H.E., J.M. Kassimo, K. Steen, T.A. Vaz and I. Tvedten. Proposition de projet la peche dans l'archipel Bijagos, Guinee-Bissau. Project report, IUCN Bissau Office, 1989.

Bakun, A. The California Current, Benguela Current, and Southwestern Atlantic Shelf ecosystems: a comparative approach to identifying factors regulating biomass yields. In: K. Sherman, L.M. Alexander, and B.D. Gold, eds. Large Marine Ecosystems: Stress, Mitigation and Sustainability. Washington, DC, AAAS Press 1992.

Bakun, A. Definition of environmental variability affecting biological processes in large marine ecosystems. In: K. Sherman and L. Herander [eds] Variability and Management of Large Marine Ecosystems, Washington, DC AAAS Press, 1986: 89-107.

Bakun, A. et al. Ocean sciences in relation to living resources. Can. J. Fish. Aquat. Sci. 1980; 39(7):1059-1070.

Bakus, G.J. The selection and management of coral reef preserves. Ocean Management 1983; 8:305-316.

Ballantine, W.J. Networks of "no-take" marine reserves are practical and necessary. In: Marine Protected Areas and Sustainable Fisheries. N.L. Shackell and J.H.M. Willison, eds. Science and Management of Protected Areas Association, Wolfville, Nova Scotia, 1995: 13-20.

Ballantine, W.J. Marine reserves for New Zealand. Univ. of Auckland, Leigh Lab. Bull., 1991;25.

Barrett, C.W., G.M. Van Dyne and E.P. Odum. Stress ecology. Bioscience 1976; 26(3):192-194.

Batisse, M. Development and implementation of the biosphere reserve concept and its applicability to coastal regions. Envir. Cons. 1990; 17(2):111-116

Batisse, M. Development and implementation of the biosphere reserve concept in coastal areas. Proc. of the Workshop on Coastal Biosphere Reserves, 14-20 August 1989, San Francisco, CA. Paris, UNESCO, 1989.

Bell, P.R.F. and I. Elmetri. Ecological indicators of large scale eutrophication in the Great Barrier Reef lagoon. Ambio 1995; 24(4):208-215.

Belleville, B. Diver-funded marine parks protect reefs and tourism. Rodales Scuba Diving, Nov./Dec. 1992:36-38.

Bensted-Smith, R. and S. Cobb. Reform of protected area institutions in East Africa. Parks 1996; 5(3):3-19.

Berkes, F. Fishermen and the tragedy of the commons. Envir. Cons. 1985; 12(3):199-206

Bernacsek, G.M. Status of the fisheries sector in Africa. Rome, Food and Agricultural Organization, Fisheries Dept., Fishery and Policy Planning Division Report, 1987.

Bernal, P. and P.M. Holligan. Marine and coastal systems. Proceedings of the International Conference on an Agenda of Science for Environment and Development into the 21st Century, 24-29 Nov 1991 Vienna: Section II, Theme 8:1-10.

Bjorklund, M.I. Achievements in marine conservation: International marine parks. Env. Cons. 1974;1(3): 205-223.

Bliss-Guest, P. and A. Rodriguez. Sustainable development. Ambio 1987; 10(6):-346.

Bohnsack, J.A. The potential of marine fishery reserves for reef fish management in the U.S. Southern Atlantic. NOAA Tech. Mem/ NMFS-SEFC-261, 1990.

Bohnsack, J.A. and J.S. Ault. Management strategies to conserve marine biodiversity. Oceanography 1996; 9(1):73-82.

Boo, E. Ecotourism: Potentials and Pitfalls. World Wildlife Fund, Wash., DC, 1990.

Broadus, J.M. A special marine reserve for the Galapagos Islands. Proceedings of Coastal Zone '85, 1985.

Broadus, J.M. and A.G. Gaines. Coastal and marine area management in the Galapagos Islands. Coastal Management 1987; 15:75-88.

Brock, V.E. and R.H. Rittenburgh. Fish schooling: a possible factor in reducing predation. J. Cons., Cons. Int. Explor. Mer 1960; 25:307-317.

Brodie, P.F., D.D. Sameoto and R.W. Sheldon. Population densities of euphasiids off Nova Scotia as indicated by net samples, whale stomach contents, and sonar. Limnology and Oceanography 1978; 23:1264-1267.

Brundtland, G. Our common future. In: V. Martin, ed. For the Conservation of the Earth. Golden, Colo. Fulcrum Pres, 1988: 8-12.

Budowski, G. Tourism and environment conservation: conflict, coexistence or symbiosis? Env. & Cons. 1976; 3(1):27-31.

Burbridge, P.R., N. Dankers and J.R. Clark. Multiple use assessment for coastal management. Coastal Zone 1989; 89:33-45.

Caddy, J.F. Toward a comparative evaluation of human impacts on fishery ecosystems of enclosed and semi-enclosed seas. Rev. Fish. Sci. 1993; 1(1):57-95.

Caddy, J. F. The protection of sensitive sea areas: a perspective on the conservation of critical marine habitats of importance to marine fisheries. In: Proc. of the International Seminar on the Protection of Sensitive Sea Areas, 1990: 17-29.

Caddy, J.F. Species interactions and stock assessment–some ideas and approaches. In: C. Bas, R. Margalev and S.P. Rubies, eds. Simposio Internacional sobre des areas de afloramiento mas importantes del Oeste Africano (Cabo Blanco y Benguela). Instituto de Investigaciones Pesqueras, Barcelona, 1985.

Caddy, J.F. and G.D. Sharp. An Ecological Framework for Marine Fishery Investigations. FAO Fish. Tech. Pap. 1983; 283

Campredon, P. The Bijagos archipelago. Case study for the Workshop on Coastal Biosphere Reserves, 14-20 Aug 1989, San Francisco, CA.

Carlton, J. and A. Cohen. Alien Invasions in San Francisco Bay. USFWS Report 1996.

Carnegie Commission On Science, Technology, and Government. E Cubed: Organizing for Environment, Energy, and the Economy in the Executive Branch of the U.S. Government. Washington, DC. Task Force on Environment and Energy, 1990.

Carpenter, R.A. and J.E. Maragos. How to Assess Environmental Impacts on Tropical Islands and Coastal Areas. SPREP Training Manual. Honolulu, HI, Environment and Policy Institute Publication, Honolulu, HI, 1989.

Carr, M.H. and D.C. Reed. Conceptual issues relevant to marine harvest refuges: examples from temperate reef fishes. Can J. Aquat. Sci. 1993; 50:2019-2028.

Carter, R.W. Coastal Environments. New York, Academic Press, 1988.

Castilla, J. and R. Duran. Human exploitation from the intertidal zone of central Chile: the effects on *Concholepas concholepas* (Gastropoda). Oikos 1985; 45:391-399.

Ceballos-Lascurain, H. Tourism, ecotourism, and protected areas. Paper presented to the Commission on National Parks and Protected Areas (CNPPA), World Conservation Union (IUCN) General Assembly Meeting, 28 Nov–6 Dec, 1990, Perth, Australia.

Chapin, F.S. III, E.D. Schulze and H.A. Mooney. Biodiversity and ecosystem processes. Trends in Ecology and Evolution 1992;7(4)

Cherfas, J. The fringe of the ocean–under siege from land. Science 1990; 248

Choat, J.H. Fish feeding and the structure of benthic communities in temperate waters. Ann. Rev. Ecol. Sys. 1982; 13:423-429.

Clark, C. Clear-cut economies. The Sciences, NY Academy of Science, Winter 1988.

Clark, J.R. Management of coastal barrier biosphere reserves. Bioscience 1991; 41(5): 331-336.

Commission d'Etat du Developpement Rural (CEDR). Development integre de la zone IV: region de Bolama. Rapport de la Phase I: Etudes et propositons preliminaires. CRAD & SUCO report, 1989.

Conrad, M. Statistical and hierarchical aspects of biological organization. In: C.H. Waddington, ed. Towards a Theoretical Biology. Edinburgh University Press, Edinburgh, 1972.

Costanza, R. Developing ecological research that is relevant for achieving sustainability. Ecol. Appl. 1993; 3(4):579-581.

Costanza, R., W.M. Kemp and W.R. Boynton. Predictability, scale, and biodiversity in coastal and estuarine ecosystems: implications for management. Ambio 1993; 22(2-3):88-96.

Cousins, K. Ecotourism–Examples from the United States Coastal management Programs and marine and estuarine parks. Coastal Zone 1991; '91:1602-1610.

Craik, W., R. Kenchington and G. Kelleher. Coral reef management. In: Dubinsky, ed., Ecosystems of the World. Elsevier 1990 Vol. 25: 453-466.

Csirke, J. Recruitment of the Peruvian anchovy (*Egraulis* riggens) and its and its dependence on the adult population. Rapp. P.V. Reun. CIEM 1980;177:307-313.

Cycon, D.E. Managing fisheries in developing nations: a plea for appropriate development. Nat. Res. J. 1986;26: 1-14.

da Silva, F. and J.L. Kromer. Projet d'amelioration des techniques artisanales de transformation du poisson. Projet du development integre des Iles Bijagos. Report 1988; 635/85/W02.

Dagg, M., C. Grimes, S. Lohrenz, B. McKee, R. Twilley and W. Wiseman, Jr. Continental shelf food chains of the Northern Gulf of Mexico. In: K. Sherman, L. Alexander and B. Gold (eds). Food Chains, Yields, Models and Management of *Large Marine Ecosystems*. Boulder, Westview Press, 1991:67-106.

Dahl, A.L. The challenge of conserving and managing coral reef ecosystems. UNEP Regional Seas Rep. 1985;69: 85-87.

Davis, G.E. Designated harvest refugia: The next stage of marine fishery management in California. CalCOFI Rep. 1989; 30.

Davis, G.E. and J.W. Dodrill. Marine parks and sanctuaries for spiny lobster fisheries management. Proc. of the Gulf and Caribbean Fisheries Institute 1989; 32:197-207.

Dayton, P.K. Scaling, disturbance, and dynamics: Stability of benthic marine communities. In: T. Agardy, ed. The Science of Conservation in the Coastal Zone. Proceedings of the IVth World Conference on Parks and Protected Areas. 1993 IUCN Marine Conservation & Development Report, Gland, Switzerland.

Dayton, P.K. The structure and regulation of some South American kelp communities. Ecol. Monogr. 1985; 55:447-468.

Dayton, P.K. Competition, disturbance, and community organization: the provision and subsequent utilization of space in a rocky intertidal community. Ecol. Monogr. 1971; 41:351-389.

Dayton, P., R. Hofman, S. Thrush, and T. Agardy. Environmental effects of fishing. Aquatic Conservation: Marine and Freshwater Ecosystems 1995; 5:205-232.

Dayton, P.K. and M.J. Tegner. Bottoms beneath troubled waters: benthic impacts of the 1982-1984 El Nino in the temperate zone. In: P. Glynn, ed. Global Consequences of the 1982-1983 El Nino Southern Oscillation. Amsterdam, Elsevier Press, 1989:457-481.

Dayton, P.K. and M.J. Tegner. The importance of scale in community ecology: A kelp forest example with terrestrial analogs. In: P.W. Price, ed. A New Ecology: Novel Approaches to Interactive Systems, 1984.

de Fontaubert, C., D. Downes and T.S. Agardy. Protecting Marine and Coastal Biodiversity and Living Resources under the Convention on Biological Diversity. Washington, DC, Center for International Environmental Law, World Wildlife Fund and IUCN Publication, Washington, DC, 1996.

de Groot, R. Functions of Nature. Amsterdam, Wolters-Noordhoff, 1992.

Dearden, P. Protected areas and the boundary model: Mears Island and Pacific Rim National Park. Geographica 1988; 32(3):256-265.

Denny, M.W. and M.F. Shibata. Consequences of surf-zone turbulence for settlement and external fertilization. Am. Nat. 1989;117: 838-840.

diCastri, F. and T. Younes. Ecosystem function of biological diversity. Biology International Special Issue 1989; 22.

Diegues, A.C. Application of the biosphere reserve concept to coastal and marine areas: Case Study #8. Proceedings of the Workshop on Coastal Biosphere Reserves, 14-20 Aug 1989, San Francisco, CA, UNESCO.

Dixon, J.A., L.F. Scura, and T. Van't Hof. Meeting ecological and economic goals: marine parks in the Caribbean. Ambio 1993; 22(2-3):117-125.

DGFC/CECI/IUCN. Projet de zone de conservation des tortues marines de l'archipel des Bijagos. Projet planification cotiere, 1991.

Djohani, R.H. Patterns of spatial distribution, diversity and cover of corals in Palau Seribu National Park: implications for the design of core coral sanctuaries. Proc. of the third IOC-WESTPAC Conference on the 'Sustainability of Marine Environment' 22-26 Nov 1994, Bali, Indonesia.

Domain, F. Rapport des campagnes de Chalutages du N.O. Andre Nizery an large des cotes de Guinee-Bissau. Institut de Recherche Agronomique de Guinee, Ministre Francais de la Cooperation, 1988.

Doumenge, F. Human interactions in coastal and marine areas: present day conflicts in coastal resource use. Proc. on the Workshop of the Biosphere Reserve Concept to Coastal Areas, 14-20 August 1989, San Francisco, CA.

Earle, S. Sea Change. New York, G.P. Putnam's Sons, 1995.

Eichbaum, W.M., M.P. Crosby, M.T. Agardy, and S.A. Laskin. The role of marine and coastal protected areas in the conservation and sustainable use of biological diversity. Oceanography 1996; 9(1):60-70.

Eichbaum, W.M. and B.B. Bernstein. Current issues in environmental management: a case study of southern California's marine monitoring system. Coastal Management 1990; 18:433-445.

Elder, D. and J. Pernetta. Oceans: World Conservation Atlas. London, Mitchell Beazley Publishers, 1991.

Eldredge, M. The last wild place: marine reserves and reef fish. Washington, DC, Center for Marine Conservation, Washington, DC, 1994.

Esping, L.E. The establishment of marine reserves. In: Proceedings of the International seminar on the Protection of Sensitive Sea Areas 1990: 378-386.

Ferreras, J. Protected coastal areas: Samana Bay. Proceedings of a Workshop on management and planning of protected tropical coastal areas, Santo Domingo, Dominican Republic, 1987.

Fiske, S.J. Sociocultural aspects of establishing marine protected areas. Ocean and Coastal Management 1992; 18:25-46.

Fogarty, M.J., M.P. Sissenwine and E.B. Cohen. Recruitment variability and the dynamics of exploited marine populations. Trends in Ecology and Evolution 1991; 6(8):241-245.

Fonds Africain de Development. Rapport d'evaluation: projet de developpement de la peche artisanal avancee, Guinee-Bissau. Dept. de l'Agriculture et du Developpement Rural II, 1990.

Fontana, A. and J. Weber. Apercu de la situation de la peche maritime Senegalaise. Centre de recherches oceanographiques de Dakar-Thiaoye. Dakar, Senegal, 1982.

Frontier, S. Diversity and structure in aquatic systems. Oceanogr. Mar. Biol. Ann. Rev. 1985; 23:253-312.

Fuentes, E.R. Scientific research and sustainable development. Ecol. Appl. 1993; 3(4):576-577.

Gaines, S.D. and M.D. Bertness. Dispersal of juveniles and variable recruitment in sessile marine species. Nature 1992; 360:579-580.

Gaudian, G. and M. Richmond. Mafia Island Marine Park Project. The People's Trust for Endangered Species. London, Imperial College, London, 1990.

GESAMP. The State of the Marine Environment. Oxford, Blackwell Scientific Publications, 1990.

Gilpin, M. and M. Soule. Minimum viable populations: processes of species extinction. In: M. Soule, ed. Conservation Biology: The Science of Scarcity and Diversity. Ann Arbor, U. Mich. Press, 1986: 19-34.

Gjerde, K. and D. Ong. Marine and coastal biodiversity conservation. A report for the Worldwide Fund for Nature, Gland, Switzerland, 1995.

Goeden, G.B. Intensive fishing and a 'keystone' predator species: ingredients for community instability. Biol. Cons. 1982; 22:273-281.

Goldemberg, J., T.B. Johansson, A.K.N. Reddy, and R.H. Williams. Energy for a Sustainable World. Washington, DC World Resources Institute, 1989.

Gosselink, J.G. and L.C. Lee. Cumulative impact assessment in bottomland hardwood forests. Center for Wetland Resources, Louisiana State University, 1987. LSU-CEI-86-09.

Grassle, J.F. and N. Maciolek. Deep-sea species richness: regional and local diversity estimates from quantitative bottom samples. Am. Nat. 1992; 139(2): 313-341.

Great Barrier Reef Marine Park Authority (GBRMPA). Rezoning plan for the Cairns Section. Townsville, Australia, GBRMPA, 1992.

Gregg, W. Draft guidelines for biosphere reserves and regional MAB programs in the U.S. Unpublished document, U.S. Park Service, Wash., DC, 1989.

Gubbay, S. Marine Protected Areas: Principles and Techniques for Management. London, Chapman and Hall, 1995.

Gubbay, S. Management of marine protected areas in the UK: lessons from statutory and voluntary approaches. Aquat. Conservation: Marine and Freshwater Ecosystems 1993; 3:269-280.

Gubbay, S. Using sites of special scientific interest to conserve seashores for their marine biological interest. A report for the World Wide Fund for Nature from the Marine Conservation Society, London, England, 1989.

Gulland, J.A. Food chain studies and some problems in world fisheries. In: J. Steele, ed. Marine Food Chains. Edinburgh, Oliver and Boyd Press, 1970: 296-315.

Gulland, J.A. and S. Garcia. Observed patterns in multispecies fisheries. In: R. May, ed. Exploitation of Marine Communities: Report of the Dahlem Workshop. 1-6 April 1984, Berlin. Springer Verlag Life Sciences Research Report 1984; 32:155-190.

Hackman, A. Inuit create a whale sanctuary. In: E. Kemp, ed. The Law of the Mother: Protecting Indigenous Peoples in Protected Areas. San Fransisco, CA, Sierra Club Books, 1993:211-217.

Halbert, C.L. How adaptive is adaptive management? Implementing adaptive management in Washington State and British Columbia. Reviews in Fisheries Science 1993; 1(3),261-283.

Haney, J.C. Seabird affinities for Gulf Stream front eddies: Responses of mobile marine consumers to episodic upwelling. J. Mar. Res. 1986; 44:361-384.

Hardin, G. Paramount positions in ecological economics. In: The Science of Sustainability. R. Costanza, ed. Environmental Economics, 1991: 47-57.

Hardin, G. The tragedy of the commons. Science 1966; 162: 1243-1248.

Hatcher, B.G., R.E. Johannes, and A.I. Robertson. Review of research relevant to the conservation of shallow tropical marine ecosystems. Oceanogr. Mar. Biol. Ann. Rev. 1989; 27: 337-414.

Havens, K. Scale and structure in natural food webs. Science 1992; 257:1107-1109.

Hayden, B.P., R.D. Dueser, J.T. Callahan, and H.H. Shugart. Long term research at the Virginia Coast Reserve. Bioscience 1991; 41(5):310-325.

Hilborn, R. and D. Ludwig. The limits of applied ecological research. Ecol. Appl. 19933(4):550-552.

Hoagland, P., Y. Kaoru and J.M. Broadus. A methodological review of net benefit evaluation for marine reserves. World Bank Environment Dept., Pollution and Environmental Economics Division. Env. Econ. Ser. 1996; 26.

Hobson, E.S. Feeding patterns among tropical reef fishes. Am. Sci. 1975; 63:382-392.

Hobson, E.S. Feeding relationships of teleostean fishes on coral reefs in Kona, Hawaii. NOAA/NMFA Fish Bull. 1974; 72(4):915-1031.

Hofman, R. Cetacean entanglement in fishing gear. Mammal. Review 1990; 20:53-64.

Holden, C. Multidisciplinary look at a finite world. Science 1990;249: 18-19.

Holling, C.S. Investing in research for sustainability. Ecol. Appl. 1993; 3(4):552-555.

Holling, C.S. Adaptive Environmental Assessment and Management. New York, John Wiley and Sons, 1988.

Holme, N.A. Fluctuations in the benthos of the western English Channel. Oceanologica Acta SP, 1983: 121-124.

Holt, S.J. Recruitment in marine populations. Trends in Ecology and Evolution 1990; 5(7): 231.

Hughes, E.L. Environmental protection in national marine parks. University of New Brunswick Law Journal 1992; Vol./Tome 41: 41-76.

Humphrey, S.R. and B.R. Smith. A balanced approach to conservation. Cons. Biol. 1990;4(4):341-343.

Hutchings, P.A. Biological destruction of coral reefs: A review. Coral Reefs 1986; 4(4): 239-252.

Iles de Paix. Proposition de projet la peche dans l'archipel Bijagos Guinee-Bissau. ASBL, 1989 Report.

International Center for Ocean Development (ICOD). Project proposal: cooperative initiatives for integrated marine resources management in the archipelago of the Bijagos. Unpublished report, 1990.

International Council for the Exploration of the Sea (ICES). Report of the study group on ecosystem effects of fishing activity. Unpublished report of the Study Group of ICES, 1992.

International Union for the Conservation of Nature and Natural Resources (IUCN). Guidelines for Protected Area Management Categories. IUCN Gland Switzerland, 1994.

International Union for the Conservation of Nature and Natural Resources (IUCN). Parks for Life: Report of the IVth World Parks Congress on National Parks and Protected Areas. IUCN, Gland, Switzerland, 1993.

International Union for the Conservation of Nature and Natural Resources (IUCN). Conservation du millieu et utilisation durable des resources naturelles dans la zone cotiere de la Guinee-Bissau. Rapport d'Activite Decembre 1989-Novembre 1990.

International Union for the Conservation of Nature and Natural Resources (IUCN). Planificao costeira da Guine-Bissau. Proposta Peliminar. Bissau, Guinee-Bissau, 1990.

International Union for the Conservation of Nature and Natural Resources (IUCN). Conservacao e desenvolvimento da zona costeira, Guine-Bissau. Bissau, Guinee-Bissau, 1990.

International Union for the Conservation of Nature and Natural Resources (IUCN). Biodiversity in sub-Saharan Africa and its islands: Conservation, management and sustainable use. IUCN Species Survival Commission Occasional Paper 1990; 6. Gland, Switzerland.

IUCN, UNEP and WRI. Caring for the Earth. Gland, Switzerland, IUCN, 1991.

James, M.K., I.J. Dight, and J.C. Day. Application of larval dispersal models to zoning of the Great Barrier Reef Marine Park. Proc. of PACON 90, 16-20 July 1990 Tokyo.

Johannes, R.E., ed. Traditional Ecological Knowledge: A Collection of Essays. Gland, Switzerland, IUCN, 1989.

Johannes, R.E. Marine conservation in relation to traditional lifestyles of tropical artisanal fishermen. The Environmentalist 1984; 4(7).

Johannes, R.E. Traditional conservation methods and protected marine areas in Oceania. In: J. McNeely and K. Miller, eds. National Parks, Conservation and Development. Smithsonian Institution Press, Washington, DC, 1984; 344-347.

Johannes, R.E. and B.G. Hatcher. Shallow tropical marine environments. In: M. Soule, eds. Conservation Biology: The Science of Scarcity and Diversity. Sunderland, MA, Sinauer, Assoc., 1986.

Johannes, R.E. and J.W. MacFarlane. Traditional Fishing in the Torres Strait Islands. CSIRO, Hobart, Tasmania, 1991.

Johannes, R.E. and M. Riepen. Environmental, economic and social implications of the live reef fish trade in Asia and the western Pacific. Arlington, VA, The Nature Conservancy Report, 1995.

Johnson, J.P. and M.P. Wilkie. Pour un development integre des peches artisanales: du bon usage de la participation et de la planification. Field Guide 1. FAO/ DIPA, 1986.

Johnston, D.M. and E. Gold. Extended jurisdiction. The impact of UNCLOS III on coastal state practice. In: T.A. Clingan, ed. Law of the Sea: State Practice in Zones of Special Jurisdiction. Honolulu, HI, Law of the Sea Inst. U. Hawaii,1982.

Jones, E. Ecosystems, food chains, and fish yields. ICLARM Conf. Proc. 1982; 9:195-239.

Jones, P.J. A review and analysis of the objectives of marine nature reserves. Ocean and Coastal Management 1994; 24:149-178.

Kelleher, G. Identification of the Great Barrier Reef region as a particularly sensitive area. In: Proc. of the International Seminar of the Protection of Sensitive Sea Areas, 1990: 170-179.

Kelleher, G. Political and social dynamics for establishing marine protected areas. Key paper, Workshop on Coastal Biosphere Reserve, 14-20 Aug 1989, San Francisco, CA. UNESCO, 1989.

Kelleher, G., C. Bleakley and S. Wells. A Global Representative System of Marine Protected Areas. Washington, DC, World Bank, 1995(4 volumes).

Kelleher,G.B. and R.A. Kenchington. Guidelines for Establishing Marine Protected Areas. IUCN Marine Conservation and Development Report, Gland, Switzerland, 1992.

Kelleher, G. and R. Kenchington. Political and social dynamics for establishing marine protected areas. Proc. of the UNESCO/IUCN Workshop on the Application of the Biosphere Reserve Concept to Coastal Areas, San Francisco, CA, 1989.

Kenchington, R.A. and M.T. Agardy. Achieving marine conservation through biosphere reserve planning. Env. Cons. 1990; 17(1):39-44.

Kenchington, R.A. and E.T. Hudson. Coral Reef Management Handbook. Paris, France. UNESCO, 1984.

Kimball, L. The Law of the Sea: Priorities and Responsibilities in Implementing the Convention. IUCN Marine Conservation and Development Report, Gland, Switzerland, 1995.

Knecht, R.W. 1990. Towards multiple use management: issues and options. In: S.D. Halsey and R.B. Abel, eds. Coastal Ocean Space Utilization. Amsterdam, Elsevier Science Publ., 1990.

Kriwoken,L.K. The Great Barrier Reef Marine Park: an assessment of zoning methodology for Australian marine and estuarine protected areas. Maritime Studies 1987; 36:12-21.

Larkin, P.A. An epitaph for the concept of MSY. Trans. Am. Fish. Soc. 1977; 107:1-11.

Lawton, J.H. Feeble links in food webs. Nature 1992; 355:19-20.

Lees, A. Traditional ownership, development needs and protected areas in the Pacific. Parks 1994; 4(1):41-47.

Lemay, M. Projet d'appui a la gestion integree des resources marines de l'archipel des Bijagos dans le cadre de l'establissement d'une reserve de la biosphere. CEIO, Division de l'afrique de l'Ouest et de l'Ocean Indien, 1990.

Levin, S. Science and sustainability. Ecol. Appl. 1993; 3(4).

Limoges, B. Preliminary report on sea turtles in the Bijagos archipelago. Unpublished report, IUCN Bissau, Guinea Bissau, 1991.

Linden, O. Oceanographic features of importance for coastal marine biosphere reserves. SFCBR, 1989.

Livingston, R.J. Trophic organization of fishes in a coastal seagrass system. Mar. Ecol. Prog. Ser. 1982; 7:1-12.

Loder, J.W. and D.A. Greenberg. Predicted position of tidal fronts in the Gulf of Maine region. Cont. Shelf Res. 1986; 6(3):397-414.

Longhurst, A.R. Analysis of Marine Ecosystems. New York, Academic Press, 1981.

Looi, Ch'ng Kim. National Marine Parks Malaysia. Dept. of Fisheries, Kuala Lumpur, 1990.

Lopez, J.M., A.W. Stoner, J.R. Garcia and I. Garcia-Muniz. Marine food webs associated with Caribbean Island mangrove wetlands. Acta Cientifica, 1988.

Lubchenco, J. et al. The Sustainable Biosphere Initiative: an ecological research agenda. Ecology 1991; 72: 371-412.

Ludwig, D. Environmental sustainability: magic, science, and religion in natural resource management. Ecol. Appl. 1993; 3(4):558-558.

Ludwig, D., R. Hilborn, and C. Walters. Uncertainty, resource exploitation, and conservation: lessons from history. Science 1993; 260(2):17;36.

MacArthur, R.H. and E.O. Wilson. An equilibrium theory of insular zoogeography. Evolution 1963; 17:373-387.

MacKinnon, J., K. MacKinnon, G. Child, and J. Thorsell. Managing Protected Areas in the Tropics. IUCN, Gland, Switzerland, 1986.

Mangel, M. Decision and Control in Uncertain Resource Systems. New York, Academic Press, 1985.

Mangel, M., L.M. Talbot, G.K. Meffe, M.T. Agardy, D.L. Alverson, J. Barlow, D.B. Botkin, G. Budowski, T. Clark, J. Cooke, R.H. Crozier, P.K. Dayton, D.L. Elder, C.W. Fowler, S. Funtowicz, J. Giske, R.J. Hofman, S.J. Holt, S.R. Kellert, L.A. Kimball, D. Ludwig, K. Magnusson, B.S.

Malayang, C. Mann, E.A. Norse, S.P. Northridge, W.F. Perrin, C. Perrings, R.M. Peterman, G.B. Rabb, H.A. Regier, J.E. Reynolds III, K. Sherman, M.P. Sissenwine, T.D. Smith, A. Starfield, R.J. Taylor, M.F. Tillman, C. Toft, J.R. Twiss, Jr.,J. Wilen, and T.P. Young. Principles for the conservation of wild living resources. Ecol. Appl. 1996; 6(2):338-362.

Mann, K.H. Physical oceanography, food chains, and fish stocks: a review. ICES J. Mar. Sci. 1993; 50:105-119.

Mann, K.H. Ecology of Coastal Waters: A Systems Approach. University of California Press, Berkeley, 1982.

Mann, K.H. General concepts of population dynamics and food links. In: O. Kinne, ed. Marine Ecology IV: Dynamics. Chichester, England, Wiley Interscience, 1978: 617-704.

Mann, K.H. and J.R. Lazier. Dynamics of Marine Ecosystems. Oxford, Blackwell Scientific Publications, 1991.

Markham, A. Potential impacts of climate change on ecosystems: a review of implications for policymakers and conservation biologists. Cli. Change Res. 1995 6:179-191.

Massin, J.M. Waste disposal at sea in light of sensitive marine areas concepts. In: Proceedings of the International Seminar on the protection of Sensitive Sea Areas 1990: 337-349.

McClanahan, T. Are conservationists fish bigots? Bioscience 1990; 40(1):2.

McClanahan, T. Kenyan coral reef-associated gastropod fauna–a comparison between protected and unprotected reefs. Marine Ecology Progress Series 1989; 53:11-20.

McManus, J.W. The Spratly Islands: a marine park? Ambio 1994; 23(3):181-186.

McNeely, J.A. Common property resource management or government ownership: Improving the conservation of biological resources. Intl. Rel. 1991; 10(3):211-225.

Methot, S. Etude preliminaire de l'archipel des Bijagos en vue de la creation d'une aire protegee. CECI publication, 1990.

Miller, D. Learning from the Mexican Experience: Area apportionment as a potential strategy for limiting access and promoting conservation of the Florida lobster fishery. In: K.Gimbel, ed. Limiting Access to Marine Fisheries: Keeping the Focus on Conservation. Center for Marine Conservation and World Wildlife Fund, Wash., DC, 1994.

Mooney, H.A. and O.E. Sala. Science and sustainable use. Ecol. Appl. 1993; 3(4):564-566.

Moore, J.C., P.C. de Ruiter and H.W. Hunt. Influence of productivity on the stability of real and model ecosystems. Science 1993; 261:906-907.

Mossman, R. Managing protected areas in the South Pacific: A training manual. SPREP/IUCN/NPS International Affairs joint publication, 1987.

National Academy of Sciences. Understanding Marine Biodiversity. Washington, DC, National Academy Press, 1995.

National Marine Fisheries Service (U.S. Dept. of Commerce, National Oceanic and Atmospheric Administration). The potential of marine fishery reserves for reef fish management in the U.S. Southern Atlantic. NOAA Tech. Mem. 1990; NMFS-SEFC-261. SAFMC, Plan Development Team.

National Oceanic and Atmospheric Administration. Florida Keys National Marine Sanctuary: Draft Management Plan/ Environmental Impact Statement. Washington, DC, NOAA, 1995.

National Research Council. Managing Troubled Waters: The Role of Marine Environmental Monitoring. National Academy Press. Wash., DC, 1990.

Naurois, R. Peuplements et cycles de reproduction des oiseaux de la cote occidentale d'afrique, du Cap Barbas, Sahara, Espagnol, a la frontiere de la Republique de Guinee. IVieme partie: Guinee Portugaise. Mem. Mus. Hist. Nat., Ser. A, Zoologie 1969; 56:191-251.

Newell, R.C. The biological role of detritus in the marine environment. In: M.J. Fasham, ed. Flows of Energy and Materials in Marine Ecosystems: Theory and Practice. New York, Plenum Press, 1984: 317-343.

Nihoul, J.C. and S. Djenidi. Perspectives in three-dimensional modelling of the marine system. Elsevier Oceanogr. Ser. 1987; 45.

Norse, E. Global Marine Biological Diversity. Washington, DC, Island Press, 1993.

Northridge, S. The environmental impacts of fisheries in the European community waters. A Report to the European Commission's Directorate General Environment, Nuclear Safety and Civil Protection. MRAG Ltd, 1991.

Odum, W.E. The relationship between protected coastal areas and marine fisheries genetic resources. In: J. McNeely and K. Miller,[eds.] National Parks, Conservation and Development. Smithsonian Institution Press, Washington, DC, 1984.

O'Neill, R.V., D.L. DeAngelis, J.B. Waide, and T.F.H. Allen. A Hierarchical Concept of Ecosystems. Monographs in Population Biology 1986; 23. Princeton NJ, Princeton University Press.

Organization for Economic Cooperation and Development (OECD). Guidelines for Aid Agencies on Global and Regional Aspects of the Development and Protection of the Marine and Coastal Environment. Guidelines on Aid and Development 1996 No. 8. Paris, France.

OECD. Guidelines for Aid Agencies for Improved Conservation and Sustainable Use of Tropical and Sub-tropical Wetlands. Guidelines on Aid and Environment 1996 No. 9. Paris, France.

Paine, R.T. Food web analysis through field measurement of per capita interaction strength. Nature 1992; 355:73-75.

Paine, R.T. Food webs: linkage, interaction strength and community infrastructure. J. Animal Ecology 1980; 49:667-685.

Parks Canada. Sea to Sea to Sea: Canada's National Marine Conservation Areas System Plan. Ministry of Supply and Services, Canada; 1995.

Parrish, R., A. Bakun, D. Husby and C. Nelson. Comparative climatology of selected environmental processes in relation to eastern boundary current pelagic fish reproduction. In: G. Sharp and J. Csirke [eds] Proc. of Expert Consultation to Examine Changes in Abundance and Species of Neritic Fish Resources. FAO Fish. Rep. 1983; 291(3):731-778.

Parsons, T.R. Biological coastal communities: productivity and impacts. [In] Coastal Systems Studies and Sustainable Development. Proceedings of the COMAR Interregional Scientific Conference, UNESCO, Paris, 21-25 May 1991. UNESCO Reports in Marine Science 1992; 64:27-37.

Parsons, T.R., M. Takahashi, and B. Hargrave. Biological Oceanographic Processes. Third edition. Oxford, Pergamon Press, 1984.

Pearsall, S. In: *absentia* benefits of nature preserves: A review. Envir. Cons. 1984; 11:3-10.

Peet, G. Protection of vulnerable areas in different geographic and ecological situations. In: Proceedings of the International Seminar on the Protection of Sensitive Sea Areas, 1990: 240-253.

Pernetta, J. Marine Protected Area Needs in the South Asian Seas Region: Vols. 1-5. IUCN Marine Conservation and Development Reports, Gland, Switzerland, 1993.

Pernetta, J. and D. Elder. Cross-sectoral, Integrated Coastal Area Planning: Guidelines and Principles for Coastal Area Development. IUCN Marine Conservation and Development Report, Gland Switzerland, 1993.

Peterson, M.N., ed. Diversity of Oceanic Life: An *Evaluative Review*. Center for Strategic and International Studies Significant Issues Series 1992; XIV(12).

Pimm, S.L. The Balance of Nature. Chicago, U Chicago Press, 1991.

Pimm, S.L. Communities oceans apart? Nature 1989; 339:13.

Pimm, S.L., J.H. Lawton, and J.E. Cohen. Food web patterns and their consequences. Nature 1991; 350:669-674.

Pineda, J. Predictable upwelling and the shoreward transport of planktonic larvae by internal tidal bores. Science 1991; 253:548-550.

Policansky, D. North Pacific halibut fishery management. In: National Research Council, Ecological Knowledge and Environmental Problem-Solving. Washington, DC, National Academy Press, 1986:138-149.

Polunin, N. Marine genetic resources and the potential role of protected areas in conserving them. Env. Cons. 1983; 10(1):31-41.

Post, J.C. The economic feasibility and ecological sustainability of the Bonaire Marine Park, Dutch Antilles. In: M. Munasinghe and J.McNeely, eds. Protected Area Economics and Policy. World Bank and IUCN, Washington, DC, 1994:333-338

Programme and Strategy Review Committee. African fisheries: reversing the decline. Report of the first meeting of the PSRC of the IDRC Fisheries Programme for Africa and the Middle East, 20-21 Mar 1990, Nairobi, Keyna. PSRC 1: The Nairobi Consultation.

Raines, P.S., J.M. Ridley and D. McCorry. The role of coral cay conservation in marine resource management in Belize. Presented at the World Parks Congress Workshops, 3-10 February 1992, Caracas, Venezuela.

Ray, G.C. Sustainable use of the ocean. In: Changing the Global Environment. New York, Academic Press, 1988:71-87.

Read, A. and D. Gaskin. Incidental catch of harbor porpoise by gill nets. J. Wildl. Mgmt. 1988; 52:517-523.

Reise, K. Long term changes in the macrobenthic invertebrate fauna of the Wadden Sea: are polychaetes about to take over? Neth. J. Sea Res. 1982; 16:29-36.

Robertson-Vernhes, J. Biosphere Reserves: the beginnings, the present, and future challenges. In: W. Gregg, ed. Proc of the Symp. on Biosphere Reserves. Fourth World Wilderness Congress, 14-17 Sept. 1987, Estes Park, CO, USA, 1989.

Robillard, M.J. and B. Limoges. Proposal for coastal planning. Unpublished document, CECI, Bissau, 1990.

Rosenberg, A.A., M.J. Fogarty, M.P. Sissenwine, J.R. Beddington, and J.G. Shepherd. Achieving sustainable use of renewable resources. Science 1993; 262:828-829.

Rubenstein, D.I. Science and the pursuit of a sustainable world. Ecol. Appl. 1993; 3(4):585-587.

Russ, G.R. and A.C. Alcala. Effects of intense fishing pressure on an assemblage of coral reef fishes. Mar. Ecol. Prog. Ser. 1989 56:13-27.

Saila, S.B. and J.D. Parrish. Exploitation effects upon interspecific relationships in marine ecosystems. NOAA/NMFS Fish Bull. 1972; 70(2):383-393.

Sale, P. A reply. Trends in Ecology and Evolution 1990; 5:25-27.

Salm, R.V. Ecological boundaries for coral reef preserves: principles and guidelines. Env. Cons. 1984; 11(3):209-215.

Salm, R.V. and J.R. Clark. 1984. Marine and Coastal Protected Areas: A Guide for Planners and Managers. IUCN Publication, Gland, Switzerland, 1984.

Salm, R.V. and J.A. Dobbin. Management and administration of marine protected areas. In: Proc. of the Workshop in the Application of the Biosphere Reserve Concept to Coastal Areas, 14-20 August 1989, San Francisco, CA, USA.

Salm, R.V. and A.H. Robinson. Moorings for marine parks: design and placement. Parks 1982; 7(2):21-23.

Salwasser, H. Sustainability needs more than better science. Ecol. Appl. 1993; 3(4):587-589.

Sepa-Marie, M. Profit a court terme on exploration durable des tortues marines de l'archipel des Bijagos. Unpublished document, CECI, 1990.

Sherman, K. Monitoring and assessment of large marine ecosystems: A global and regional perspective. In: McKenzie, ed. Ecological Indicators. Amsterdam, Elsevier Press, 1992.

Sherman, K. The large marine ecosystem concept: research and management strategy for living marine resources. Ecol. Appl. 1991; 1(4):349-360.

Sherman, K. Sustainability of resources in large marine ecosystems. In: K. Sherman, L. Alexander, and B. Gold, eds. Food Chains, Yields, Models, and Management of Large Marine Ecosystems. New York, Westview Press, 1991: 1-34.

Sherman, K. Can large marine ecosystems be managed for optimum yield? In: K. Sherman and L. Alexander, eds. Variability and Management of Large Marine Ecosystems. AAAS Selected Symposium 1986; 99: 263-267.

Sherman, K., L.M. Alexander, and B. Gold, eds. Food Chains, Yields, Models, and Management of Large Marine Ecosystems. New York, Westview Press, 1990.

Simenstad, C.A., J.A. Estes, and K. Kenyon. Aleuts, otters and alternate stable-state communities. Science 1978; 200:403-411.

Sinclair, M. Marine Populations. Seattle, WA, U. Washington Press, 1988.

Skud, B.E. Dominance in fishes: the relation between environment and abundance. Science 1982; 216: 144-149.

Slocombe, D.S. Implementing ecosystem-based management. Bioscience 1993; 43(9):612-622.

Smayda, T. Global epidemic of noxious phytoplankton blooms and food chain consequences in large ecosystems. In: K. Sherman, L. Alexander, and B. Gold, eds.Food Chains, Yields, Models, and Management of Large Marine Ecosystems,1991: 275 -308.

Smith, A.H. and F. Berkes. Solutions to the 'tragedy of the commons': sea-urchin management in St. Lucia, West Indies. Env. Cons. 1991; 18(2):131-136.

Smith, H.D. The regional bases of sea use management. Ocean Shore. Mgmt. 1991; 15: 273-282.

Smith, P.E. Biological effects of ocean variability: time and space scales of biological response. Rapp. Cons. Int. Explor. Mer. 1978; 173:117-127.

Sobel, J. Conserving marine biological diversity through marine protected areas: a global challenge. Oceanus 1993; 36(3):22-23.

Socolow, R. Achieving sustainable development that is mindful of human imperfection. Ecol. Appl. 1993; 3(4):581-583.

Sorensen, J.C., S.T. McCreary, and M.J. Hershman. Coasts: Institutional Arrangements for Management of Coastal Resources. Renewable Resources Information Series, Coastal Management Pub. 1, 1984.

Soule, M.E. and D. Simberloff. What do genetics and ecology tell us about the design of nature reserves? Biol. Cons. 1986; 35:19-40.

Starr, M., J.H. Himmelman, J.C. Therriault. Direct coupling of marine invertebrate spawning with phytoplankton blooms. Science 1990; 247:1071-1074.

Steele, J.H. Marine functional diversity. Bioscience 1991; 41(7):470-474.

Steele, J.H. A comparison of terrestrial and marine ecological systems. Nature 1985; 313:355-358.

Steele, J.H. The Structure of Marine Ecosystems. Cambridge, MA, Harvard University Press, 1974.

Strathmann, R.R. Why life histories evolve differently in the sea. Amer. Zool. 1990; 30:197-207.

Tegner, M.J. and P.K. Dayton. Sea urchin recruitment patterns and implications for commercial fishing. Science 1977; 196:324-326.

Thorne-Miller, B. and J. Catena. The Living Ocean : Understanding and Protecting Marine Biodiversity. Wash., DC, Island Press, 1991.

Ticco, P.C. An analysis of the use of marine protected areas to preserve and enhance marine biological diversity–A case study approach. Ph.D. Dissertation, U. Delaware, Newark, DE, 1995.

Tisdell, C. Sustainable development: differing perspectives of ecologists and economists, and relevance to LDCs. World Development 1988;

16(3):373-384.

Todd, J. An ecological economic order. Fifth Annual E.F. Schumacher Lecture, Cambridge, MA: Harvard University, 1985.

Towle, E.L. and C.S. Rogers. Case study on the Virgin Islands Biosphere Reserve. Case study, Workshop on Coastal Biosphere Reserves, 14-20 Aug 1989, San Francisco, CA. UNESCO.

United Nations Educational, Scientific and Cultural Organization (UNESCO). Action Plan for Biosphere Reserves. UNESCO, Paris, France, 1984.

United Nations Educational Scientific and Cultural Organization (UNESCO). Biosphere Reserves: The Seville strategy and statutory framework for the world network. UNESCO, Paris, France, 1996.

United Nations Environment Programme (UNEP). Guidelines for the selection, establishment, management and notification of information on marine and coastal protected areas in the Mediterranean. Regional Activity Centre for Specially Protected Areas, Tunis, Tunisia, 1987.

United Nations Environment Programme. Protocol Concerning Mediterranean Specially Protected Areas. Coordinating Unit for the Mediterranean Action Plan, Tunis, Tunisia, 1986.

Uravitch, J. A. Strategies for management and regulations of particularly sensitive sea areas: experience of the National Marine Sanctuaries Program of the U.S. Proceedings of the International Seminar of the Protection of Sensitive Sea Areas, Malmo, Sweden, 1990.

van Claasen, D., ed. The application of digital remote sensing techniques in coral reef, oceanographic and estuarine studies. UNESCO, Paris, 1985.

van't Hof. Saba Marine Park: a proposal for integrated marine resource management in Saba. Netherlands Antilles National Parks Foundation (STINAPA), 1985.

Vermeij, G.J. When biotas meet: understanding biotic interchange. Science 1991; 253:1099-1104.

Voskresensky, K. 1990. Intergovernmental work in the maritime vulnerable areas. In: Proceedings of the International Seminar on the Protection of Sensitive Sea Areas, 1991:6-16.

Walters, C. Adaptive Management of Renewable Resources. New York, Macmillan Publ., 1987.

Weber, P. Abandoned Seas: Reversing the Decline of the Oceans. Worldwatch Paper 166, Wash., DC: Worldwatch Institute 1993.

White, A.T., L.Z. Hale, I. Reynard and L. Cortesi. Collaborative and Community Based Management of Coral Reefs. West Hartford, CT, Kumarian Press, 1995.

White, A.T. and V. P. Palaganas. Philippine Tubbataha Reef National Marine Park: status, management issues, and proposed plan. Env. Cons. 1991; 18(2): 148-157;136.

Witman, J. and K. Sebens. Regional variation in fish predation intensity: a historical perspective in the Gulf of Maine. Oecologia 1992; 90:305-315.

Woodley, S. Management of water quality in the Great Barrier Reef Marine

Park. Water Science and Technology 1989; 21(2):31-38.

Woodmansee, R.G. Ecosystem processes and global change. In: T. Rooswall, R. Woodmansee and P. Kisser, eds. Scales and Global Change. J. Wiley & Sons; 1988: 11-27.

World Bank. National Environmental Strategies: Learning from Experience. World Bank Environment Department: Land, Water and Natural Habitats Division. Wash., DC, 1995.

World Bank. World Bank Participation Sourcebook. Environment Department, Social Policy and Resettlement Policy Division, Participation Series Working Paper 1996; 19.

World Resources Institute. The State of the World. WRI, Washington, DC, 1996.

Yurick, D.B. International networking of marine sanctuaries. Oceanus 1988; 31(1):82-87.

# INDEX

Printed and bound by CPI Group (UK) Ltd, Croydon, CR0 4YY

03/10/2024

01040313-0016